Published by:

Gibson Applied Engineering and Technology, Inc (GATE)

Houston, TX

www.GATEINC.com

ISBN-13: 978-1548424817

ISBN-10: 1548424811

Edition 1

Author

Howard Duhon, P.E.

hduhon@gateinc.com

www.gateinc.com

TABLE OF CONTENTS

PREFACE

HAZOP – A Critically Important Process

Everyone remembers their first HAZOP. Mine was in 1991. It was the HAZOP of one of my designs. I was intrigued. I was humbled by the results.

The HAZOP is probably the most commonly used method for process hazard analysis (PHA). Needless to say, it's important to get it right. Yet every HAZOP I've attended and facilitated over the last 25 years has left something to be desired. After each, I've pondered how it could be done more effectively.

This book is the result of 25 years of pondering.

What's Wrong with HAZOPs?

There are several main reasons why typical HAZOPs are not as effective as they could be including:

1. Tunnel vision. Typical HAZOPs focus on one equipment item at a time (equipment-based nodes). The focus on small nodes obscures the big picture. Systems issues might be missed; people not already familiar with the process may not be aware of the interconnections between equipment items and between systems. It is particularly difficult to follow recycle streams and process/process heat transfer.

2. HAZOPs are supposed to identify operability issues, but there is no explicit consideration of operability in a typical HAZOP and no effective way to identify operability issues. Success at identifying operability issues depends more on having the right people in the room (experienced operators) than on the process itself.

3. Guideword overlap – The guidewords for a typical HAZOP are Flow, Pressure, Temperature, and Level. But flow deviations cause most pressure, temperature and level deviations. Discussing all four of the guidewords duplicates effort and causes tedium.

4. HAZOP reports are difficult to read and therefore are not used for much after the HAZOP.

5. Most people don't enjoy HAZOPs and since they often run for days or weeks at a time, the resulting tedium impacts the effectiveness of the process.

6. Risk assessment in HAZOPs is often ambiguous and weak. Ineffective risk assessment generates two problems:

    a. Significant hazards might be missed.

b. Can result in a large number of spurious recommendations that occupy the design team's time, are difficult to close and contribute little to improving process safety.

The GATE HAZOP Method

This book describes the GATE HAZOP Method, the Stream-Based HAZOP. The GATE Method addresses all the problems listed above. Significant features of the method include:

1. Stream-based nodes for the FLOW discussions. A stream-based node follows an identifiable stream from its inception to a logical termination point. The graphic below is a single stream-based node for a produced oil stream from the reservoir to the Dry Oil Tank.

2. Two-part process: The stream-based nodes are ideal for the FLOW discussion, but equipment-based nodes (typical HAZOP nodes) are necessary for the PRESSURE, TEMPERATURE, and LEVEL safeguarding discussions.

3. Use of LOPA rules to simplify and improve risk assessment.

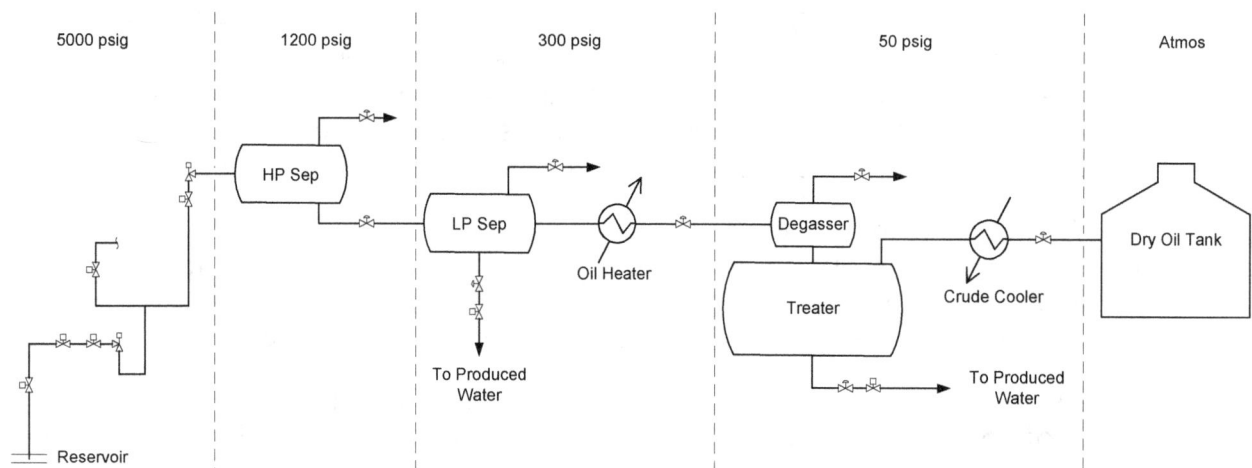

Oil Production Stream-based Node (one 'stream' from Reservoir to the Dry Oil Tank)

Sources

Much of what is in this book has been previously published by the author in SPE and AIChE papers [1,2,3,4].

SECTION I – DISCUSSION OF THE METHOD

CHAPTER 1: HAZOP PROCESS OVERVIEW

We assume that you have attended one or more HAZOPs and understand the process. This chapter is provided as a refresher for completeness. Each HAZOP is a bit different. HAZOP facilitators put their own spin on the process. So what is described here may be a little different than what you have experienced.

The HAZOP Process

The HAZOP process is a structured brainstorming process designed to identify process risks and operability issues in a design.

The HAZOP session is a team effort. There will be a number of people in the room from the design team whose role is to describe and 'defend' the design. There must also be people in the room not directly involved in the design whose job is to challenge the design.

The team should consist of:

- A facilitator who is not involved in the design and who guides the HAZOP team through the process

- Process Engineer and/or others responsible for and involved in the design effort

- Process Engineer and/or other subject matter experts not involved in the design

- Health, Safety, Environmental, Regulator specialist(s)

- Operations

- Commercial partners

- Vendor Rep (major equipment items)

The HAZOP process, as typically applied, is described below:

1. First, the facility is divided into nodes.

    a. A node is typically limited to one major equipment item and associated piping, control systems, safeguarding systems, etc. In this book, we describe these nodes as equipment-based nodes.

2. Next, the responsible design engineer describes the node, including:

    a. Design Objectives

    b. Operating and Design Conditions

    c. Process Control

d. Safeguarding

3. Then the facilitator guides the discussion of the node using the relevant guidewords selected (Flow, Pressure, Temperature, Level, etc.) and relevant deviations selected (High, Low, More, Different, etc.)

    a. For each relevant deviation-guideword pair, such as 'High Pressure', the HAZOP team:

        i. Identifies the causes of 'deviation-guideword'.

        ii. Identifies the consequence(s) of the deviation.

        iii. If the consequence is significant, identifies what safeguards/mitigations exist.

        iv. If the existing safeguards are deemed to be inadequate, makes a recommendation for improvement or records a 'finding'.

        Note: It is important to avoid getting the HAZOP team too involved in solving the identified problem. For this reason, some facilitators are rigorous in limiting the report to 'findings' (identifying a problem) and avoid recommending solutions.

4. After the session, the HAZOP facilitator issues a report. The report should include a section summarizing the recommendations/findings and a documentation section that captures the entire HAZOP session discussion.

Following the HAZOP, the design team will evaluate the recommendations and make appropriate changes to the design.

It may be necessary to reconvene the HAZOP team to review the changes made to close the HAZOP recommendations.

That was a very high level description. For a detailed explanation the definitive reference document for the HAZOP process is IEC 61882, Hazard and operability studies (HAZOP Studies) – Application Guide [14].

Requirements for a Successful HAZOP

There are three key requirements for achieving an effective HAZOP:

1. The process design being HAZOPed must be a competent design with relatively few design errors/issues.

2. The HAZOP session must be effectively conducted such that the majority of design errors are discovered without generating many spurious and needless recommendations.

3. Effective follow-up efforts must be conducted to evaluate the HAZOP recommendations, and implement those that are justified.

The requirement for a competent design prior to the HAZOP needs some elaboration. One might think that the number of problems in a design is immaterial. It is possible to conduct a HAZOP on a design at any level of design maturity. Some HAZOPs are intentionally conducted on immature designs in order to identify issues requiring further study. But when HAZOPing a detailed design, it is important that the design be relatively error free. There are at least three good reasons for this:

1. **Probability.** Suppose that a competent HAZOP team will find 95% of issues/errors in a design. If they find five problems, we can feel confident that they didn't miss any. If they find 100 problems, odds are they missed some.

2. **Tedium.** When the team finds too many problems they may begin to burn out and the effectiveness of the HAZOP decreases over time.

3. **Closeout of the recommendations.** Closing out HAZOP issues can be difficult, time consuming and error prone. The design team will be under time pressure, will likely resist making some changes and may suffer tunnel vision when evaluating some of the recommendations. The "solutions" to HAZOP recommendations frequently cause problems worse that the ones they solve.

Strengths of HAZOPs

HAZOPs are successful in part because they help us to overcome some cognitive limits to learning. For example:

Anchoring [5]. Once we form an opinion we 'anchor' on that opinion. It is hard to let go. Anchoring is particularly strong once we've make a public commitment (like a set of P&IDs). Where a design team has anchored on a design concept, the HAZOP team brings a new perspective.

Structural Secrecy [6]. Project organizations are structured into multiple teams. It is hard to know what other teams are doing. Our natural tendency to support our team tends to inhibit transfer of information to other teams. Often a design team, operating in a silo, is simply unaware of the objectives of one of the stakeholders. By bringing multiple stakeholders into the discussion, HAZOPs tend to overcome structural secrecy issues.

Alternative-focused Thinking [7]. Humans are action oriented. We tend to focus more attention on alternatives as opposed to objectives. Failure to effectively identify our objectives frequently results in poor decisions. HAZOPs focus attention on safety and operability objectives.

Conflicting Objectives. Different stakeholders have different, sometimes conflicting, objectives. HAZOPs can help mitigate conflicts by bringing the conflicting parties into a structured discussion.

Unactionable Plans, Groupthink, In-Group Favoritism [8,9,10]. We support our team when challenged. Further, we tend to view our team more favorably than we view others. This makes it hard to learn from others. Design teams can easily defend a design to each other. The HAZOP provides an independent set of eyes.

Lack of Motivation to Change [11]. Good ideas may not be implemented simply because of inertia. Projects aggressively manage change via management of change (MOC) processes which intentionally make it difficult to implement changes. It takes motivation to implement a change. A design team may know that a design is flawed, but lack the motivation to change it. A HAZOP recommendation can provide the motivation.

Stress and Expertise (Naturalistic Decision Making) [12]. Time pressure impacts our decision making in predictable ways. For example, under stress, we look for satisfactory options (satisficing) vs. optimal solutions and we may suffer from tunnel vision. The expertise provided by the HAZOP team can decrease the stress of making difficult project decisions.

For these and other reasons HAZOPs have had a significant positive effect on design safety over the past three decades, but we can also identify shortcomings in the HAZOP technique and in the way it is commonly applied.

CHAPTER 2 – WHAT'S WRONG WITH HAZOPS?

Structural Limits to HAZOP Effectiveness

Tunnel Vision

Typical HAZOPs focus on one equipment item at a time, (equipment-based nodes). The focus on these small equipment-based nodes obscures the big picture which can result in:

- Systems issues might be missed.

- People that are not already familiar with the process may not be aware of the interconnections between equipment items and between systems.

- It is particularly difficult to follow recycle streams and process/process heat transfer.

Structural Secrecy

One of the HAZOP strengths noted in Chapter 1 is mitigation of structural secrecy. Structural secrecy is a term used by sociologists to describe communication limits created by the way we structure organizations.

An engineering organization is composed of multiple teams of specialists. Individuals (who are members of teams) collect raw data, evaluate it (add meaning) and then transmit summaries and inferences to other teams.

The collection and evaluation of data is not unbiased. Team membership impacts what individual team members believe, what data they collect, how they evaluate the data and what they report. Discussions within teams may be relatively open and honest, but team-to-team and team-to-management communications are likely to be highly filtered and may be designed more to achieve team objectives than to transmit facts and valid information.

And even when the information conveyed is accurate and meaningful, there will almost always be a time delay. Teams need to carefully vet the information they feed to other teams and up to management.

The HAZOP decreases structural secrecy by bringing members from multiple teams together (sometimes for the first time).

But two features of HAZOPs limit their effectiveness in this respect:

1. The focus on small nodes.

   This minimizes the potential for team-team sharing. If you are looking at a single separator will the HAZOP team notice that pressure control occurs twice on the gas stream – once at the separator and once at the compressor.

2. The practice of dividing HAZOPs along contract boundaries contributes to structural secrecy.

   The worst examples of structural secrecy occur at the interfaces between contracts. We recognize that this is an area of concern and most projects now apply significant resources to managing these interfaces.

   But a real communication opportunity is missed when HAZOPs stop at the contract boundary.

Guideword Overlap

Flow, pressure, temperature, and level are usually all discussed in the HAZOP. But flow deviations cause most pressure, temperature and level deviations. If you rigorously do all the flow, pressure, temperature and level deviation discussions you will perform the HAZOP twice.

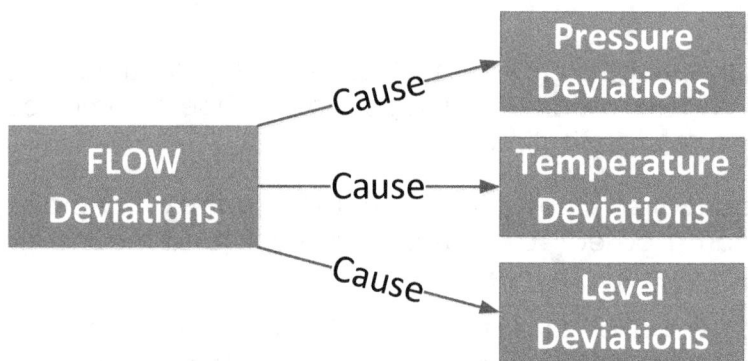

Many facilitators start with the FLOW deviations discussion. Effectively they do the whole HAZOP on FLOW. That is not a problem conceptually, but in practice it is a problem.

Consider a process with Nodes A, B, and C discussed in that order:

- High Flow in Node A causes a High Pressure in Node B.

- Low Flow in Node C also causes a High Pressure in Node B.

Resulting problems include:

1. The Node B pressure discussion happens partially in Nodes A and C and under the FLOW guideword.

2. The HAZOP documentation reflects this. If you want to know what the HAZOP team discussed about over-pressure in Node B you'll have to search through multiple nodes and multiple guidewords.

3. There is also a problem of ensuring completeness. If you talk about the causes of high pressure in Node B in three or more different places instead of all at once, how do you know and how can you show that you have covered them all?

4. You may not accurately assess the probability of high pressure if you don't consider all causes at once. The safeguards provided may be adequate for any single cause of high pressure, but may be inadequate if there are multiple causes.

5. There is a resultant lack of focus on the integrity of the safeguards. Every time the team identifies a cause of high pressure, the safeguards are listed (actually copied from a previous discussion). If you talk about the pressure safeguards on a vessel 8 different times, no one is paying attention for the last 7 of those. Wouldn't it be better to focus serious attention on the safeguards one time only?

6. And, worst of all, discussions of pressure, temperature, and level (the things we actually care about) are diminished. Under "High Pressure" the HAZOP record is often full of notes such as "See High Flow in Node Y", and "See Low Flow in Node X".

HAZOP Reports Are Difficult to Read

Guideword overlap, as discussed above, results in HAZOP reports that are difficult to read. Because they are difficult to read and follow, they are not used for much after the HAZOP.

No Explicit Focus on Operability

The 'OP' in 'HAZOP' means that we are supposed to use the process to identify operability issues, but there is nothing explicit in the HAZOP process that helps us do that. Success depends mainly on having the right people, such as experienced operators, in the room.

Some HAZOPs include guidewords chosen to spark operability discussions, such as startup, shutdown, etc. But here again, the small nodes limit meaningful discussion. It is difficult to discuss operation of a single piece of equipment – many operability issues emerge at a system level.

Tedium

Most people don't enjoy HAZOPs and since HAZOPs often run for days or weeks at a time, the tedium impacts the effectiveness of the study.

Think about this: you gather 6 to a dozen subject matter experts into a room to critique someone else's design. What could be more fun than that? And yet the typical HAZOP process makes it tedious.

Risk Assessment in HAZOPs is Often Ambiguous & Weak

The qualitative risk matrices used for risk assessment in HAZOPs require consequence and probability judgements that are difficult for the HAZOP team to make. In many HAZOPs, every vessel over-pressure 'results' in a fatality! That is a serious over-estimation of the consequence in most cases.

Ineffective risk assessment generates two problems:

1. Significant hazards might be missed (or deemed not credible).

2. Can result in a large number of spurious recommendations that occupy the design team's time and are difficult to close.

CHAPTER 3 – STREAM-BASED NODES

The central and distinguishing idea of the GATE HAZOP Process is the stream-based node. Stream-based nodes are based on streams rather than equipment items. Your intuition may suggest that a stream is even smaller than an equipment item node.

This sketch illustrates what we mean by a stream-based node.

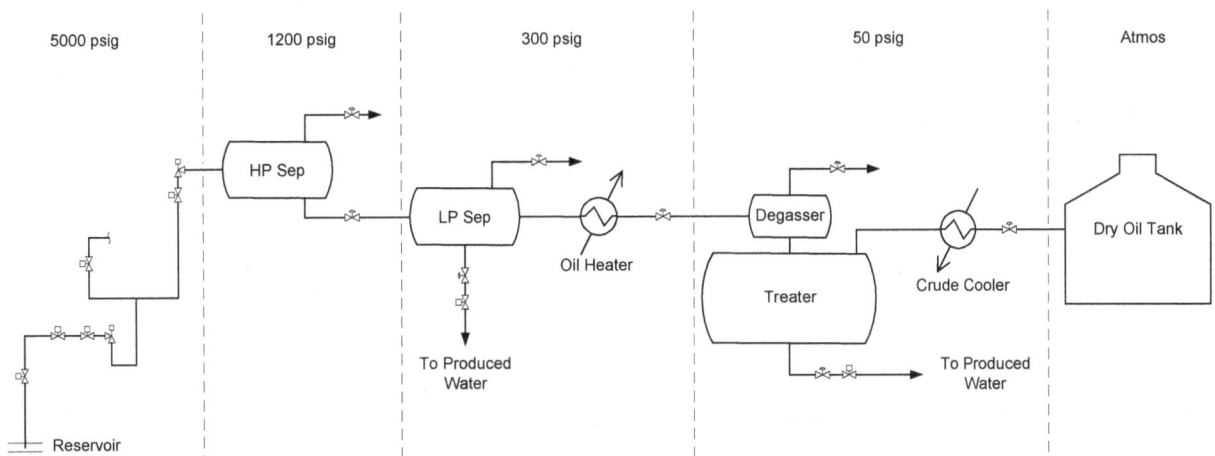

Figure 3.1: Oil Production Node

In this node, oil from the reservoir sand face travels through the separation and oil treating systems to the Dry Oil Tank. This is one stream-based node.

The stream is changed along the way: it is heated and cooled, gas is flashed off, water is separated out, but it is still one 'stream'.

Stop the flow anywhere and you stop it everywhere.

Why do it this way? Why stream-based nodes?

In a typical HAZOP, the oil production stream-based node described above would be divided into 6 to 9 equipment-based nodes.

The stream-based node solves the tunnel vision problem of small nodes.

Stream-based nodes are much better for FLOW deviations discussions. Change the flow anywhere in the node and you change it everywhere, or at least everywhere downstream.

Structural secrecy is decreased. In the node pictured in Figure 3.1, there are at least 3 design teams and major contractors involved – subsurface, subsea, and topsides.

HAZOPs divided according to contract boundaries may miss interface issues.

Flow does not respect contract boundaries. Process risk does not respect contract boundaries. Stream-based nodes do not respect contract boundaries.

The stream-based node also enables a discussion of operating procedures. We can talk about how the node in Figure 3.1 will be started up, shutdown, commissioned, operated with equipment out-of-service, etc. It is usually not very meaningful to ask how an individual vessel will be started up, but it makes perfect sense to discuss how the oil separation train will be started up.

With stream-based nodes, we can get a sense of how disturbances propagate through the process. Recycles and split flows are easier to see and evaluate. This is particularly useful when considering compression systems where the recycle often breaches three or mode equipment-based nodes.

Stream-based nodes allow more effective evaluation of process control and process complexity.

Simple questions like sampling and chemical injection are easier to visualize and evaluate at the stream-based node level.

Spec breaks are easier to evaluate.

It is easier to visualize the impact of out of service equipment.

Flow Deviation Guidewords

Our standard list of FLOW deviation guidewords (Table 3.1) suggests the breadth of discussions topics possible when you are discussing FLOW in a stream-based node.

Table 3.1 FLOW Deviations Checklist

- More Flow, Capacity, Bottlenecks
- Less Flow, No Flow, Turndown
- Flow Measurement (sizing, turndown),
- Process Control
- Complexity (Physical, Organizational)
- Different Flow (Composition, Two-phase, Chemical Injection, Solids)
- Corrosion, Erosion
- Sampling
- Spec Breaks
- Recycle, bypass, split flow
- Commissioning
- Startup
- Shutdown, Blowdown
- Maintenance (equipment out of service)

Identifying Streams

To the uninitiated it is not immediately obvious how to define stream-based nodes.

Here is an example to demonstrate stream-based node definition for the gas condensate process summarized via block diagram in Figure 3.2.

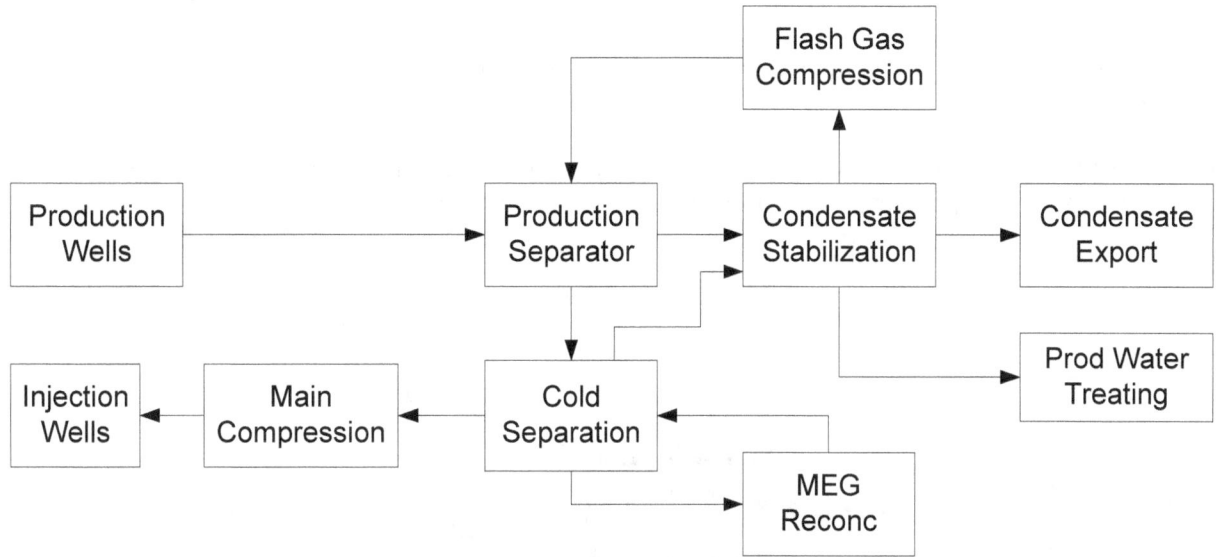

Figure 3.2: Gas Condensate Process Block Diagram

The process includes:

- Production Wells feed Inlet Separation.

- Gas from Separation goes to a Cold Separation process for additional condensate recovery.

- Lean gas is compressed and injected via injection wells.

- A MEG system provided hydrate protection in the cold sections

- The condensate is stabilized and exported.

- Produced water is treated for disposal

So, what are the streams? A few are identified and discussed below.

Node 1: Gas Stream

Figure 3.3 shows the first node: the gas stream.

In this node, gas is produced (via Production Wells) and then treated to recover condensate (liquid). The first liquid recovery step is Production Separation where liquid is recovered by dropping the stream pressure. The gas flows from Production Separation to Cold Separation, where additional liquid is recovered via cooling. Following Cold Separation, the gas is compressed (via main Compression) and re-injected into the Reservoir.

From sand-face to sand-face this node covered 42 P&IDs, but it is one stream. Stop the flow anywhere and you stop it everywhere (or divert it to flare).

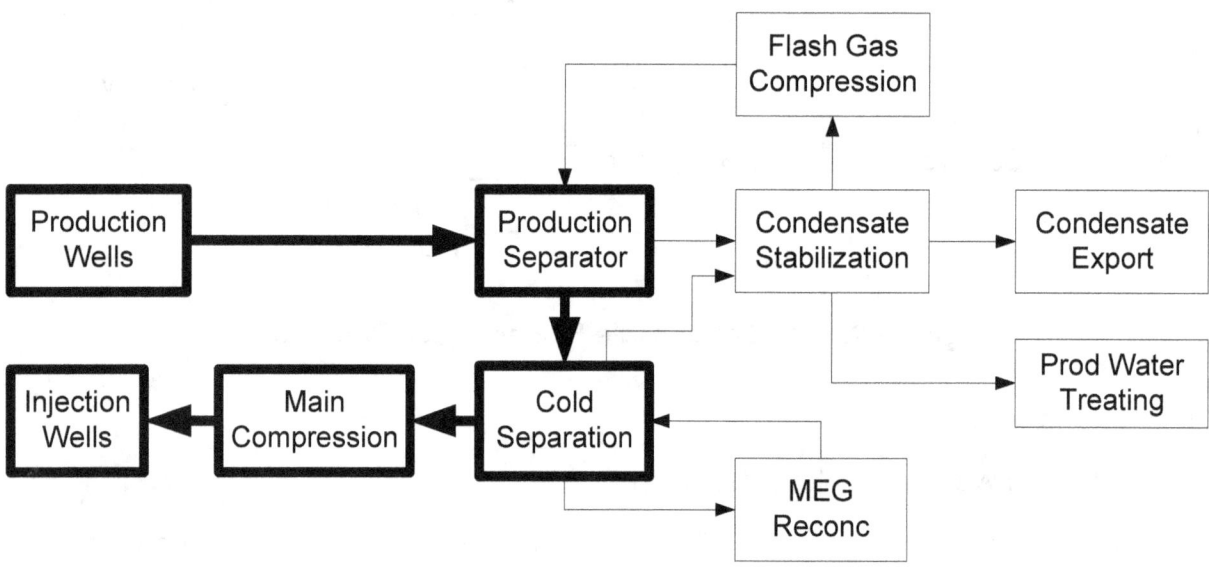

Figure 3.3 - Node 1: Gas Production through Reinjection

Node 2: Condensate

A second logical 'stream' is the condensate recovered from the gas stream in Node 1.

The condensate stream has two origins, Separation and Cold Separation. Those two streams join together and then are processed (stabilized) in the Condensate Stabilization System.

The stopping point for the Condensate node is arbitrary. It could be extended to cover Condensate Export (Stabilized Condensate stream). In this HAZOP, we chose to treat the Stabilized Condensate as a separate node for two reasons:

1. There is a Condensate Surge Tank that provides a buffer between the Condensate and Stabilized Condensate streams.

2. The Condensate Export System contains high pressure pumps; the Stabilization System contains a distillation column. Different subject matter experts are required for effective discussion of the two nodes.

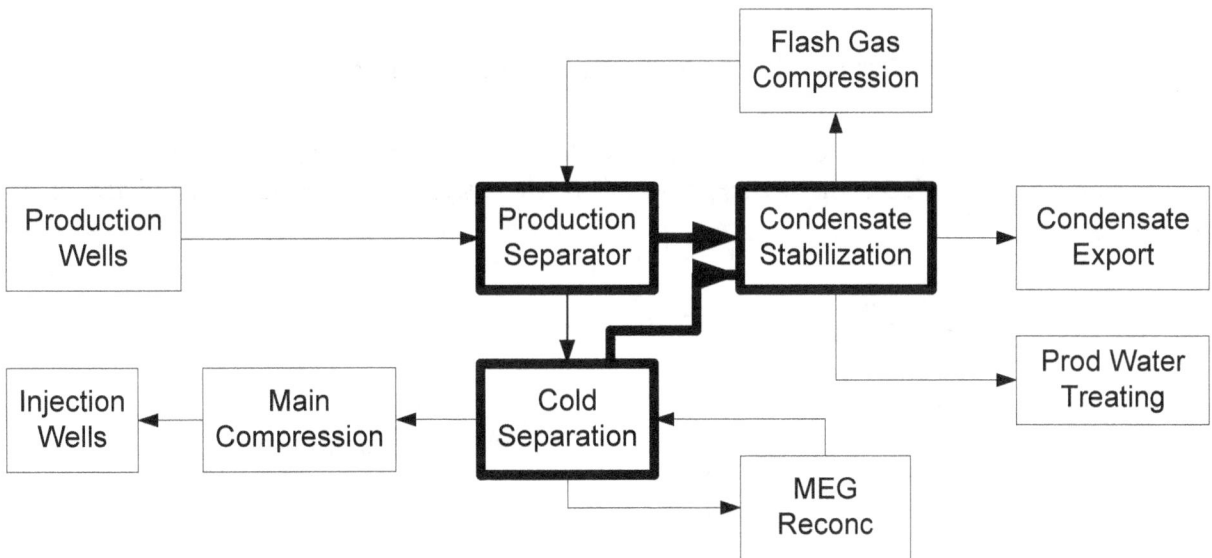

Figure 3.4 - Node 2: Condensate

Node 3: Stabilized Condensate (Condensate Export System)

Stabilized Condensate flows through the Condensate Export System.

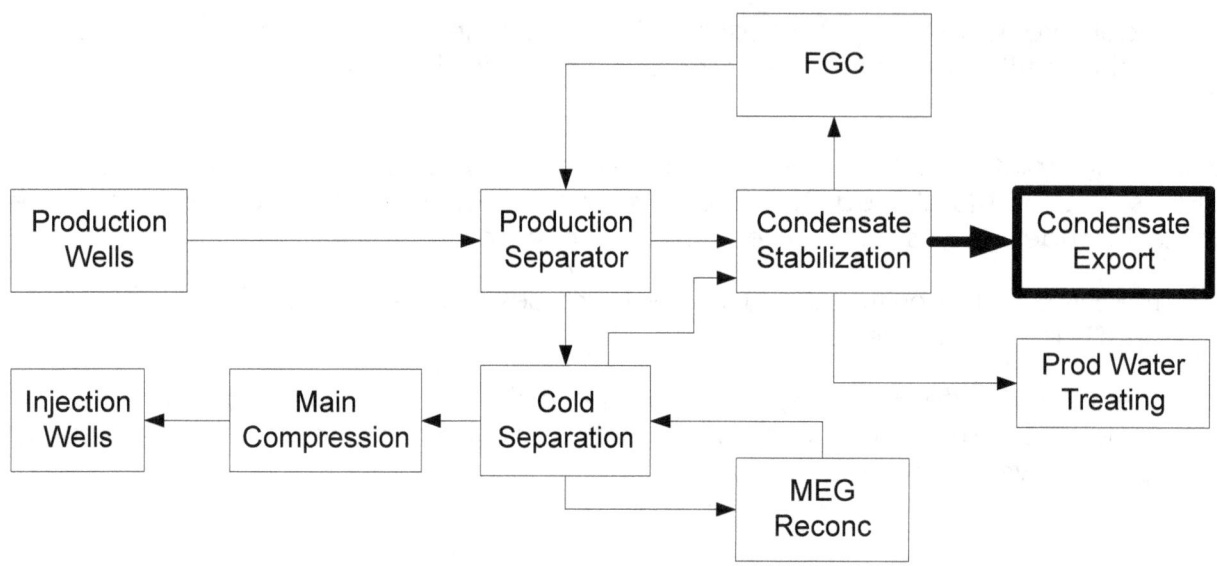

Figure 3.5 - Node 3: Condensate Export

Here it is useful to talk about node boundaries. Where does an export system node end? In a typical HAZOP, the Condensate Export System discussion would end at the Export Line boarding valve (BSDV).

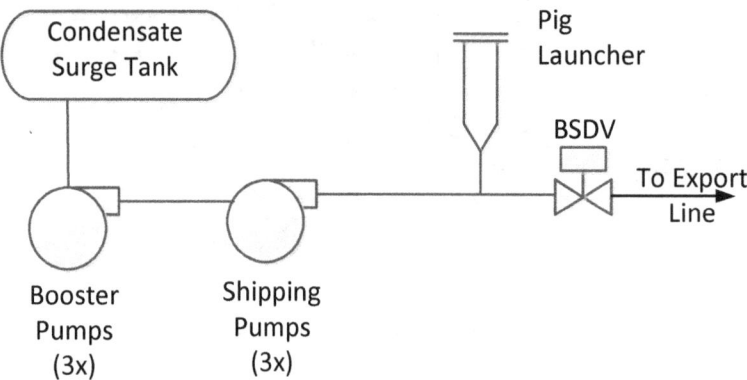

Figure 3.6 - Stabilized Condensate System Schematic

But the BSDV cannot be the end of a stream-based node! The stream continues beyond that point.

In this particular process the exported condensate flows via the export line to an FPSO 20 km away. The logical end point for the Stabilized Condensate stream-based node is the FPSO Crude Storage Tanks.

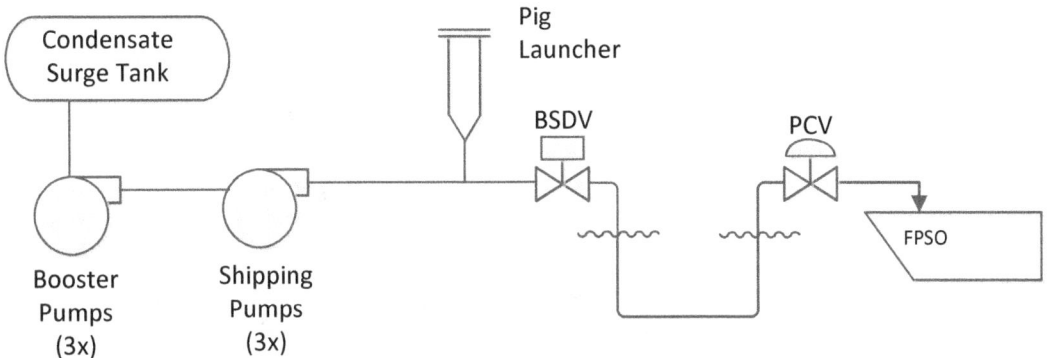

Figure 3.7 - Stabilized Condensate Node – Extended to FPSO Crude Storage Tanks

For this HAZOP, we obtained P&IDs from the FPSO operator, defined a stream-based node that extended to the Condensate Storage Tanks on the FPSO, and invited the FPSO operator to send a representative to the HAZOP. The resulting discussions highlighted significant issues on the FPSO side caused by lack of communication during the design phase.

Needless to say, these issues would not have been identified in a typical HAZOP, and they were not identified in the HAZOP of modification on the FPSO that had already occurred.

Note on HAZOPing a System That Crosses Contract Boundaries

One of the great benefits of a stream-based HAZOP is that it gives a big-picture view and avoids the tunnel vision of small nodes.

But streams cross contract boundaries and HAZOPs are usually done separately for separate design contracts.

There are reasons for dividing HAZOPs according to contract boundaries; in particular, a whole different set of design engineers is involved and different subject matter experts will likely be required.

A solution is to conduct only the FLOW discussions for the stream-based node with all contractors present, and leave the PRESSURE, TEMPERATURE, and LEVEL discussions for the contract-specific HAZOPs (see Chapter 4).

A relatively small group of engineers from each team will be needed for the FLOW discussion.

In the example given above, a single representative from the FPSO operator provided enough input for the FLOW discussion of the Stabilized Condensate Node.

Other Nodes

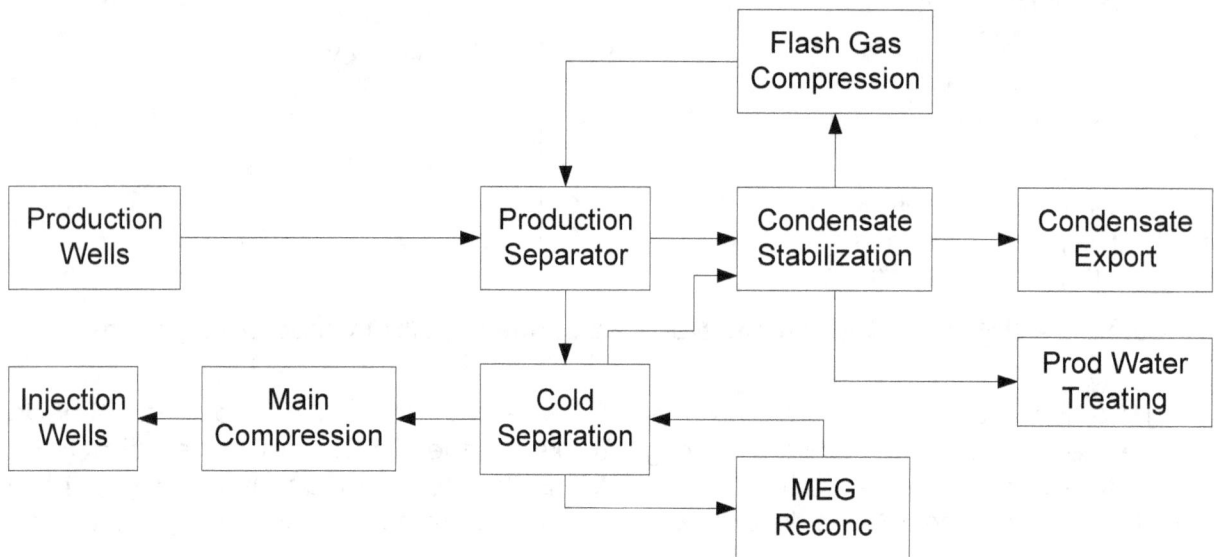

Figure 3.8 - Gas Condensate Process Block Diagram

We've discussed just 3 of major stream-based nodes. With a scan of the block diagram, the other nodes should be apparent. They include:

Flashed Gas - Gas from Condensate Stabilization is compressed and returned to Production Separation (at which point it joins Node 1 gas on its way to Cold Separation and Reinjection.

MEG - This is a recycle stream with rich MEG from Cold Separation and Lean MEG returned the Cold Separation.

Produced Water - Produced water treated for overboard discharge.

And that's it. Six process stream-based nodes are required for this HAZOP.

There will be several other utility system stream-based nodes including Closed Drain, Open Drain, Flares, etc.

Node Sketches

Because the stream-based nodes often extend over several P&IDs, we draw node sketches that summarize the node and provide much of the information needed by the HAZOP team. Here is a simplified node sketch for a portion of the Stabilized Condensate node.

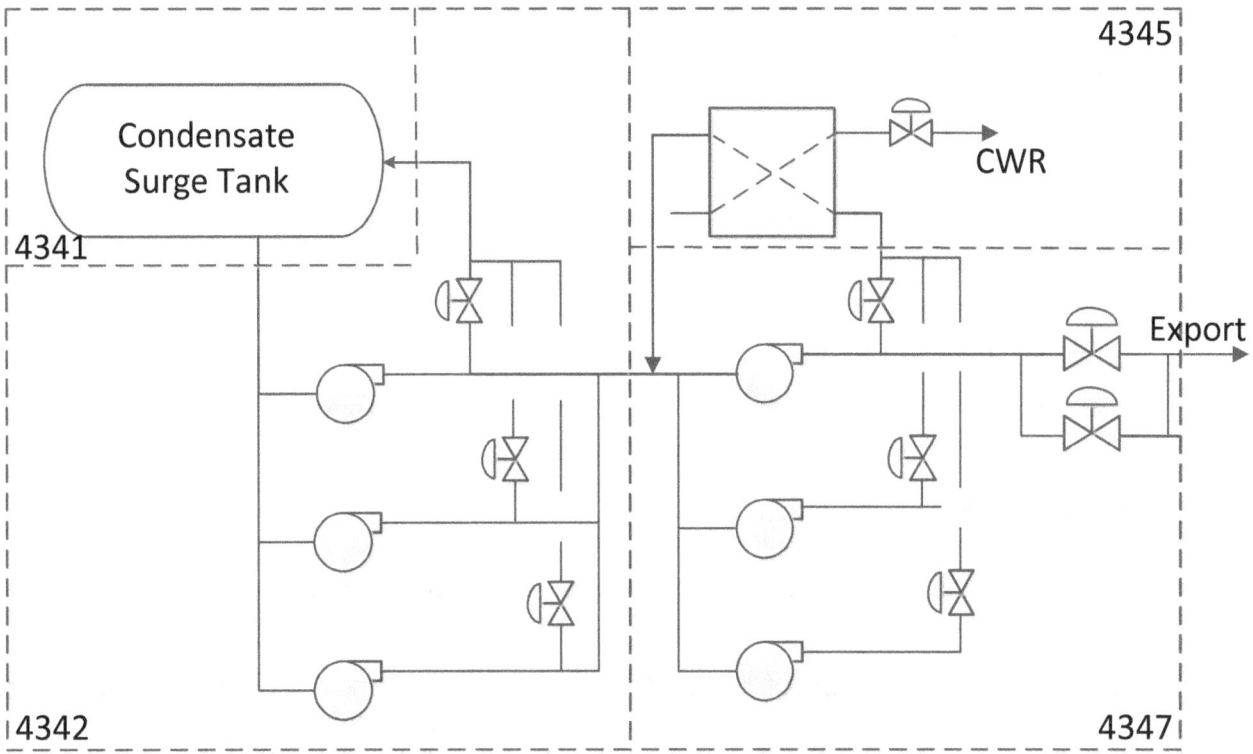

Figure 3.9 - Condensate Export System Node Sketch

Fig 3.9 illustrates 2 important features of node sketches:

1. It shows P&ID boundaries. This makes it very easy to reference the P&IDs for details.

2. It shows all equipment items including all three Booster Pumps and all three Pipeline Pumps. In this case, this is useful for showing that, while there are three Pipeline Pumps, there is only one recycle cooler. Node sketches show all major equipment items including redundant items. **It is not OK to show 1 item as "typical of 3".**

Figure 3.9 is a highly simplified node sketch. A proper node sketch contains:

- All Major Equipment Items (show all items including redundant items)
- P&ID boundaries with associated document numbers
- Major lines
- All actuated valves
- Occasionally key manual valves
- Spec Breaks
- Simplified control loops (where the objective of a loop is obvious it is adequate to just show the control valve)
- Safeguarding switches and transmitters
- PSVs
- Chemical injection points
- Sampling points
- Minor equipment needed for the HAZOP discussion (strainers, filters, static mixers, etc.)

Glycol System Node Sketch Example

Node sketches are particularly useful for systems that include recycle and process/process heat exchange.

Here is a simplified node sketch of the MEG node (MEG Reconcentration System). Most people cannot keep straight in their heads what the hot/hot exchanger does or the cold/cold exchanger, or where the condensate separator is in the process.

Drawn effectively as a single node, with a focus on the flow path, the glycol system is much easier to make sense of.

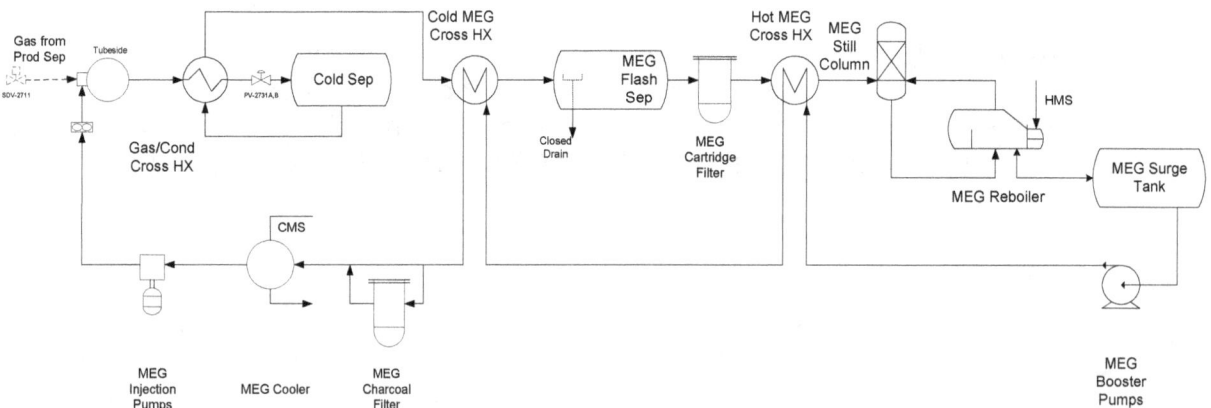

Figure 3.10 - MEG (Glycol Reconcentration System) Simplified Node Sketch

CHAPTER 4: TWO-PART PROCESS

The GATE HAZOP process has two parts:

1. FLOW discussions based on the stream-based nodes.

2. Followed by PRESSURE, TEMEPRATURE and LEVEL discussions based on the equipment-based nodes.

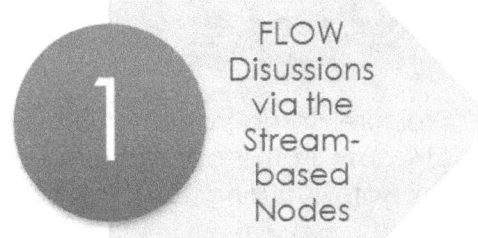 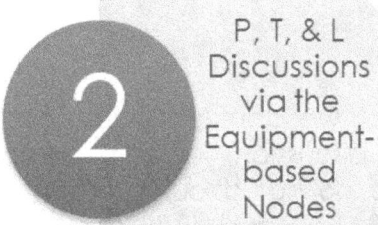

Equipment-based Nodes for Pressure, Temperature, Level Discussions

The stream-based node is an excellent vehicle for the FLOW discussions.

But stream-based nodes don't work for the PRESSURE, TEMPERATURE and LEVEL discussions.

Pressure, temperature and level control and safeguarding are all done primarily at the equipment level. For these discussions, we use equipment-based nodes.

Equipment-based nodes are what we commonly see in a HAZOP – one equipment item or a very few related items make up a node. The most important consideration in defining an equipment-based node is the location of spec breaks. The equipment-based node should generally not have multiple pressure ratings. It may also be important to have the same temperature rating throughout the equipment-based node.

Part 1: FLOW Discussions

In the GATE HAZOP, the FLOW discussions are conducted first for all stream-based nodes. These FLOW discussions identify almost all of the causes of pressure, temperature and level deviations.

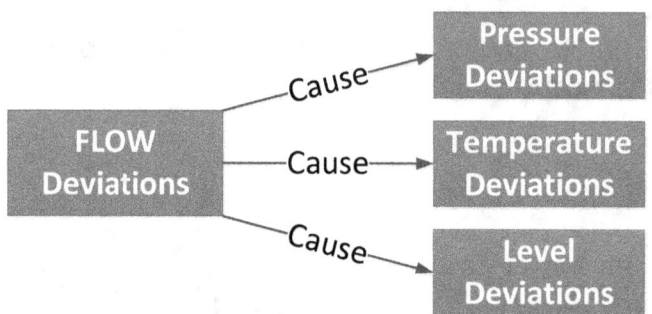

IMPORTANT: When we identify a cause of either Pressure, Temperature or Level deviation during the FLOW discussion, we simply record that cause in the appropriate equipment-based node. We do not discuss consequences or safeguarding until we conduct the PRESSURE, TEMPERATURE and LEVEL discussions in Part 2.

As we developed the stream-based nodes for the gas condensate process above, you may have noticed that some equipment items were included in more than one node. That is common. It is important to conduct all the FLOW discussions impacting an equipment-based node prior to conducting the PRESSURE, TEMPERATURE, and LEVEL discussions for that node.

Part 2: PRESSURE, TEMPERATURE, LEVEL Discussions

An equipment item will generally be part of more than one stream-based node. As we have stated previously:

> ***An equipment-based node is evaluated only after all relevant***
>
> ***stream-based nodes have been discussed.***

An important part of the stream-based discussions is identification of the causes of pressure, temperature and level deviations. Recall that these deviation causes are simply recorded (not fully discussed) during the FLOW discussions.

So, as we begin the PRESSURE, TEMPERATURE and LEVEL discussions for an equipment-based node we have already identified almost all of the pressure, level, and temperature deviation causes for that node.

The equipment-based node discussion is started with the list of causes identified during the stream-based node discussions followed by the question:

> **"are there any other causes of _____?"**

Now, confident that all deviation causes have been identified, we can discuss the frequency of occurrence, consequences and safeguarding for the equipment-based node <u>only one time</u>.

In addition to simplifying the deviation discussions, this approach also effectively organizes the HAZOP record. For example: all of the pressure deviation discussion are conducted and documented under the PRESSURE guideword and within the appropriate equipment-based node.

Figure 4.1 outlines the steps taken during the GATE HAZOP process as discussed above.

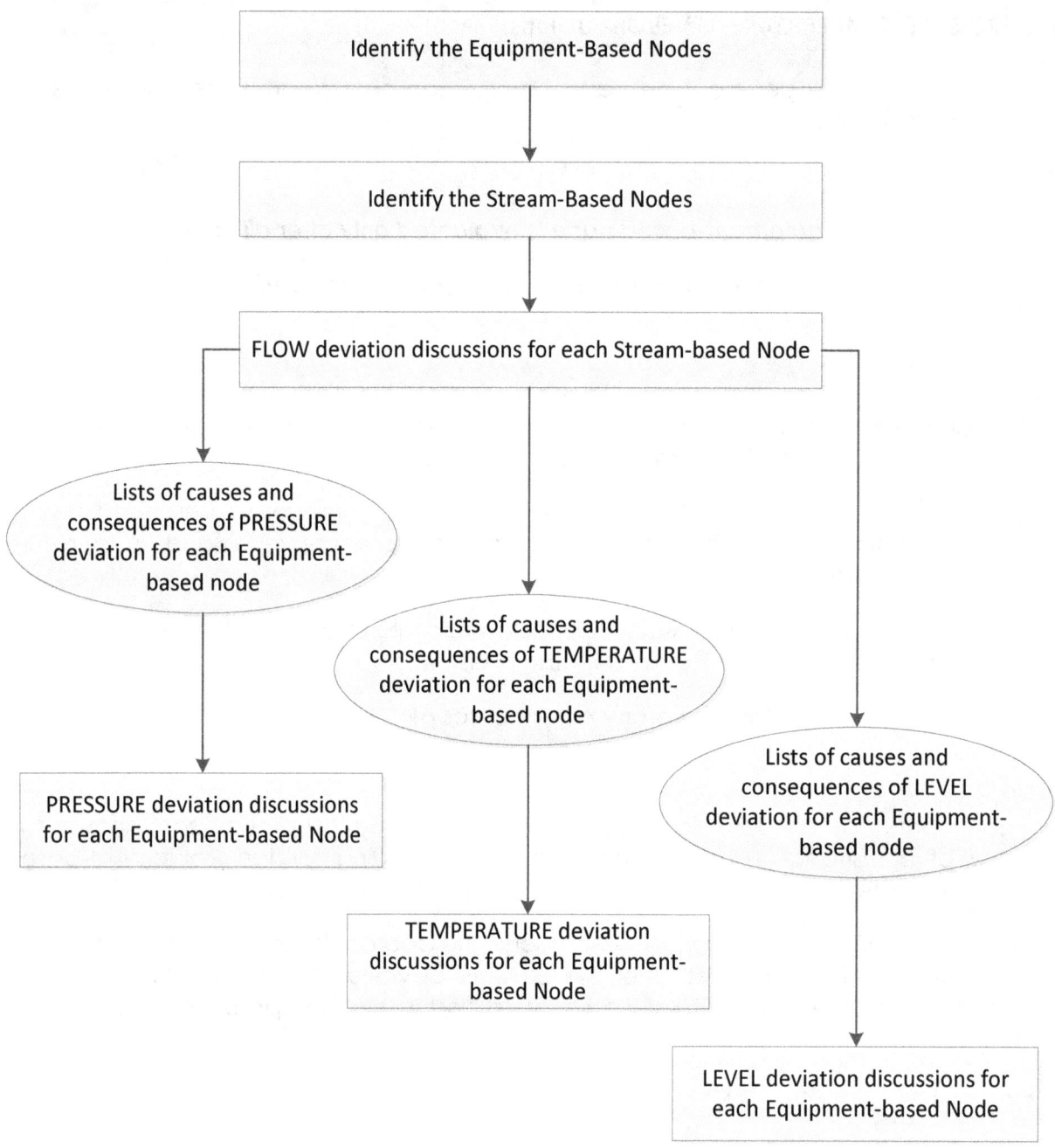

Figure 4.1: GATE HAZOP Process Summary

CHAPTER 5: RISK ASSESSMENT IN HAZOPS

Risk assessments in HAZOPs are typically performed using a simple risk matrix (Figure 5.1). In the risk matrix approach, participants make judgments as to the potential severity and the likelihood of an event. The combination of severity and likelihood indicates the risk. The risk matrix will often be color coded with green areas (OK), red areas (Unacceptable) and yellow areas (improvement suggested, subject to ALARP).

	A	B	C	D	E
5 – Catastrophic Multiple fatalities, Loss of Platform, Long-lasting environmental damage				Red	Red
4 – Major Single fatality, Significant equipment damage, Lasting environmental damage	Green				Red
3 – Severe Severe lost time injury, Significant equipment damage, Release requires cleanup but is not lasting	Green	Green			
2 – Minor Reportable injury, Equipment maintenance repair, Reportable release – no cleanup	Green	Green	Green		
1 – Slight First Aid Case, Shutdown without damage, Flaring within permitted amount	Green	Green	Green	Green	
	A Never heard of this in the industry	**B** Happens in the industry	**C** Happens in our company	**D** Likely to happen on our platform	**E** Likely to happen multiple times on our platform

Figure 5.1: Typical Qualitative Risk Matrix

This approach is problematic for a number of reasons:

1. The judgment of consequence severity is difficult and is often ambiguous. Any identified scenario could play out in multiple ways often with dramatically different outcome severities.

2. Estimation of the frequency or likelihood of the event is also difficult; especially so if the frequency of the mitigated event (rather than the unmitigated event) is to be estimated.

3. The simple green, yellow, red bands do not provide sufficient resolution for ranking scenarios. For instance, the yellow band may span multiple orders of magnitude.

This chapter presents a more rigorous and repeatable approach to making the severity and frequency judgments that is also simpler and quicker. The method is, in effect, a simplified layer of protection analysis (LOPA).

The approach also yields a HAZOP record that is more easily used for a subsequent LOPA study.

Using the Risk Ranking Matrix

Using the risk matrix requires two judgments;

1. Consequence Severity (Y-axis)

2. Event Frequency (X-axis)

Both of these judgments can be difficult. Let's look a scenario and the corresponding consequence and frequency judgements.

Consequence Judgment

Consequence severity judgment is difficult, in part, because scenarios can progress in a number of different ways. Consider the node drawn in Figure 5.2, a MEG (mono-ethylene glycol) tank and pump. On 'Low Level' the tank goes empty and the pump runs dry.

HAZOP Team A might consider this a pump damage scenario and rate it as an Asset Loss Category 2.

HAZOP Team B might identify potential for a seal leak, with blanket gas from the vessel vented to atmosphere and igniting with a possible fatality. They rate this as a Major Safety Hazard, Category 4.

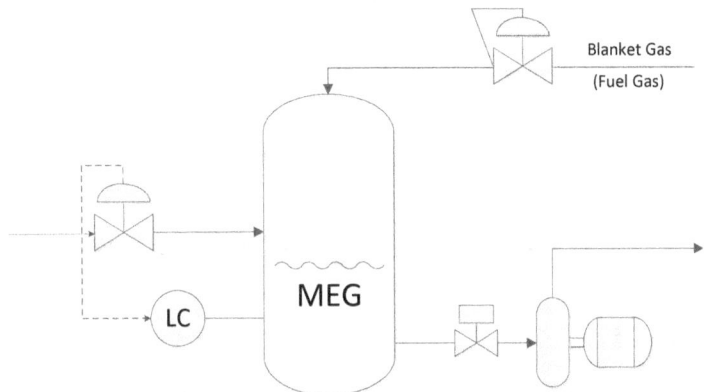

Figure 5.2: MEG Tank and Pump

The Frequency Judgment

Frequency judgments may be even more difficult and ambiguous. The guidance provided in Figure 5.1 seems reasonable at first glance. Experienced professionals can distinguish between incidents that rarely happen and incidents that happen more frequently.

A problem with such judgments is that we rarely know the full circumstances of previous incidents, and we tend to judge frequency based on the initiating event rather than on full description of the scenario.

Returning to the scenario above featuring low level in the MEG Tank:

Team A (consequence is pump damage, Category 2 – Minor) may assess the frequency as D, "likely to happen on our platform". This results in a 'yellow' ranking

If Team B (consequence is gas release and fatality, Category 4 – Major) also ranks the frequency as likely to happen, they will also get a 'yellow'.

5 – Catastrophic				Red	Red
4 – Major	Green			*Team B*	Red
3 – Severe	Green	Green			
2 – Minor	Green	Green	Green	*Team A*	
1 – Slight	Green	Green	Green	Green	
	A	**B**	**C**	**D**	**E**

Figure 5.3: Risk Rankings by Teams A and B

There are a couple of large problems here:

1. The two 'yellow' results will look the same in a report, but they are not nearly identical. One is up against a 'green' cell and could be considered 'almost green' and the other is up against a 'red' cell and should be considered 'almost red'.

2. The frequency judgement for the more serious consequence should be much lower than the frequency judgement for the lesser consequence.

Participants may judge the frequency as fairly high, because we have all seen level control systems fail, but the frequency of the level control failing is not the frequency that we need to estimate. We must be estimating the frequency of occurrence <u>of the selected consequence.</u>

If the consequence selected is a fatality, the estimated frequency must consider many contributions:

1. The frequency of the tank going empty (without safeguards) may be fairly high - once in 10 years is the typical assumed frequency for a control loop failure.

2. An independent low level shutdown will decrease this outcome frequency, perhaps by one order of magnitude. The frequency of the level loop failure and low-level shutdown failure occurring simultaneously is then 1/100 years.

3. The frequency of the tank going empty plus the pump seal being damaged may be somewhat lower.

4. The frequency of the tank going empty, plus seal damage, plus blanket gas blowby, plus ignition of the gas cloud is much lower.

5. The frequency of all of the above plus someone being exposed and killed by the flash fire is lower still.

In order for the risk matrix approach to be useful, the frequency judgment must be based explicitly on the consequence severity assumed and not just on the frequency of the initiating event.

Both of the judgments required by the Figure 5.1 risk matrix are difficult.

The GATE Approach to Risk Ranking

We suggest the following approach to make risk assessments in HAZOPs easier, more consistent and more accurate:

1. Generate and use a risk matrix that yields order of magnitude risk reduction targets.

2. Apply simple rules for estimation of consequence severity

3. Use initiating event frequency data, conditional probability data, and probability of failure on demand (PFD) data to make the frequency judgments.

Required Risk Reduction (RRR) Matrix

Figure 5.4 is an example of a required risk reduction (RRR) matrix, calibrated to yield risk reduction targets.

The important features of this matrix are:

1. The frequency axis is a logarithmic scale, with an order of magnitude change from one column to the next.

2. The consequence axis is also a logarithmic scale with order of magnitude differences between rows.

3. The numeric cell entries represent order of magnitude risk reductions required to reach a 'target' or 'maximum acceptable' risk level.

		A	B	C	D	E	F
Catastrophic	5	1	2	3	4	5	B
Major	4	0	1	2	3	4	B
Severe	3	-1	0	1	2	3	B
Minor	2	-2	-1	0	1	2	3
Slight	1	-3	-2	-1	0	1	2
		A	B	C	D	E	F
		1/10,000 Years	1/1,000 Years	1/100 Years	1/10 Years	1/1 Years	10/1 Years

Figure 5.4 – Example Risk Reduction Target Matrix

For example, per this matrix, a scenario with a Major consequence (Row 4) with an expected frequency of 1/10000 years (Column A) represents a maximum acceptable risk. The entry in cell 4A is '0', which means that there is no required risk reduction (subject to ALARP/RAGAGEP principles).

The same scenario occurring at a frequency of 1/100 years (cell 4C) is judged unacceptable with an RRR target of 2 (i.e. 2 orders of magnitude, or a factor of 100, risk reduction is required).

Using the Risk Reduction Target Matrix in a HAZOP

To use the RRR matrix in a HAZOP, for a given hazard scenario:

1. Determine the unmitigated risk level:

    a. Determine the consequence severity for the scenario (see section titled "MAKING THE CONSEQUENCE JUDGEMENT", this chapter below)

    b. Determine the frequency of the initiating event and adjust that frequency (as necessary and appropriate) to account for enabling events, multiple required events, etc. (see section titled "MAKING THE FREQUENCY JUDGEMENT", this chapter below)

    c. Enter the matrix at the selected consequence severity row and initiating event column. This is the unmitigated risk - the cell entry provides the risk reduction target.

2. Apply existing safeguards/IPLs to determine the mitigated risk.

3. If necessary, recommend additional safeguard(s) to achieve the target risk level. Landing on a cell with a value greater than 0 is automatically a finding.

Example 1

Hazard scenario: A blocked outlet of a relatively small vessel yields overpressure high enough to cause rupture of the vessel. A PSV is installed, but no other safeguarding is provided.

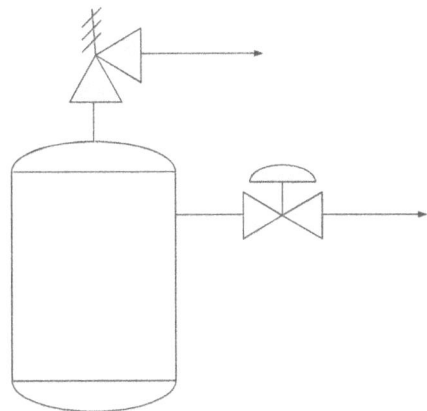

Figure 5.5: Vessel Overpressure

1. Determine the unmitigated risk:

   a. HAZOP team identifies a blocked outlet scenario with potential for fatality (Row 4, Major).

   b. The initiating event is pressure control loop failure. Per Table 5.5, control loop failure occurs 1/10 years (Column D).

   c. Enter the matrix at 4D. The required risk reduction (RRR) target is 3.

2. The only safeguard provided is the vessel PSV which provides a 2 order of magnitude improvement (see Table 5.7).

3. That leaves a 1 order of magnitude risk reduction requirement. That requirement could be provided by a safety instrumented function (SIF) as shown in Figure 5.5, or other valid independent protection layer (IPL) or via a design modification.

		A	B	C	D	E
Catastrophic	5	1	2	3	4	5
Major	4	(0) ←SIF (1) ←2 PSV (3)				4
Severe	3	0	0	1	2	3
Minor	2	0	0	0	1	2
Slight	1	0	0	0	0	1
		A	B	C	D	E
		1/10,000 Years	1/1,000 Years	1/100 Years	1/10 Years	1/1 Years

Figure 5.6: Example Use of the Required Risk Reduction Matrix

Some Advantages of this approach:

1. The first estimate made above (RRR = 3) reflects the inherent, unmitigated process risk, without safeguards. This allows judgment of inherent risk that is not usually apparent in a HAZOP. A process with a very high unmitigated RRR, say 4 or greater, should be considered unacceptable and require redesign rather than mitigation.

2. The final risk assessment reflects the mitigated risk with existing safeguards. This makes it very clear to all participants whether the risk judgment being made is the mitigated or unmitigated risk and provides clear guidance on whether the mitigated risk is acceptable.

3. If a recommendation needs to be made it can be much more precisely articulated.

4. Risk ranking is more precisely detailed into orders of magnitude risk rather than green, yellow, red.

5. Since the risks have meaningful numerical values, a comprehensive assessment of risk for the entire facility could be made by adding the individual mitigated risks.

Example 2

Recall the MEG Tank low level scenario above.

HAZOP Team A identified the consequence of low level to be pump damage (Minor, Category 2) and considered the probability to be "Likely to Happen on our Platform" which we can take to mean a frequency of 1/10 years.

HAZOP Team B identified a consequence of gas release and fatality and assumed a similar frequency of occurrence.

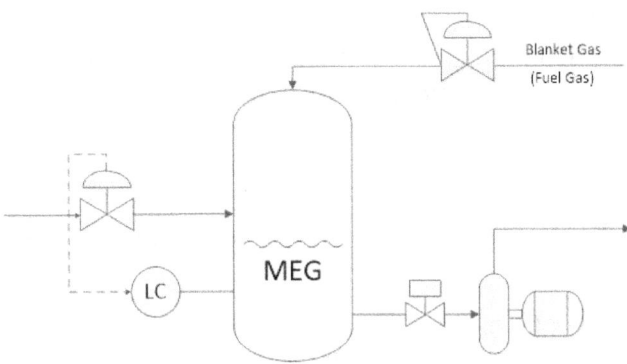

The risk management process proposed here will yield dramatically different probabilities for these two consequence assumptions:

For Consequence = Pump Damage:

> Frequency of occurrence = Frequency of control loop failure = 1/10 years.

> The SDV provides a one order of magnitude mitigation so the mitigated frequency will be 1/100 years.

For Consequence = Fatality:

> Mitigated frequency of occurrence = Freq of Loop Failure (1/10) **x** Prob of SDV failure (1/10) **x** Prob of Ignition (1/10) = 1/1000 years.

> And if it can be argued that no person will be there then multiply by another (1/10) = 1/10,000 years.

Risk Reduction Matrix Development

The first requirement of an effective matrix is that each row and column increments by one order of magnitude.

The next step is to populate the cells with numeric RRR targets. This requires that we identify an anchor point.

Note the large bold **0** in cell 4A. The source of this anchor point is:

- First, we define a "Major" event as one likely to cause a single fatality.

- The average healthy 30-year-old person has about a 1/1,000 (10^{-3}) annual probability of dying from all causes (e.g., injury, illness, etc.). Although many people are surprised when they first hear this number, it is a level of risk that we implicitly accept.

- We want the workplace to be safer than the world at large. Hence cell 4A has a '0'. It is 0 because 1/10,000 years (10^{-4}) is the <u>maximum</u> frequency at which a Major work-related event (e.g. potential fatality) is considered tolerable.

This selected anchor point is for illustration only. Another point may be selected to reflect the risk tolerance of the operating company.

		A	B	C	D	E
Catastrophic	5					
Major	4	**0**				
Severe	3					
Minor	2					
Slight	1					
		A	B	C	D	E
		1/10,000 Years	1/1,000 Years	1/100 Years	1/10 Years	1/1 Years

Figure 5.7: Matrix Anchor Point

Once having identified the anchor cell, then the rest of the matrix is easily populated since each row and column is one order of magnitude different than neighboring rows and columns.

Safety, Environmental, Economic Consequence Definitions

We can define the Consequence Categories to achieve or approximate order of magnitude steps as follows (Table 5.1):

Severity	Safety	Environmental	Cost
5 Catastrophic	Multiple Fatalities	50,000 bbls dead oil	1 Billion $US
4 Major	Single Fatality	5,000 bbls dead oil	100 Million $US
3 Severe	Serious Injury	500 bbls dead oil	10 Million $US
2 Minor	Minor Injury	50 bbls dead oil	1 Million $US
1 Slight	First Aid	5 bbls dead oil	100 Thousand $US

Table 5.1 – Consequence Definitions

Making the Consequence Judgment

There is some unavoidable uncertainty and ambiguity in consequence severity assessment. The method used should provide:

1. Guidance on identifying the range of possible consequences.

2. Agreement on selecting one or more of the identified consequences for further analysis.

3. Explicit agreement that the frequency of occurrence should be matched to the specific consequence selected.

Matching the Frequency to the Consequence

For HAZOP purposes, we are typically guided to identify the worst credible consequence, but:

- a more severe consequence, even if orders of magnitude less likely, may generate a larger RRR. And

- a less severe consequence may be much more likely to occur and hence yield a larger RRR.

Suggestion: Identify the range of possible outcomes from the most likely to most severe. Estimate the frequency of each. Choose the consequence/frequency pair that yields the largest RRR.

Consequence of Vessel Overpressure

Many HAZOP teams consider any overpressure to be a Major (4) or even Catastrophic (5) event. This is overly conservative.

The current ASME pressure vessel code provides for a design safety factor of about 3.5 for Div 1 vessels (most vessels are designed per Div 1).

Topsides piping designed to B31.3 has a safety factor of 3.0.

While exceeding the design pressure is a notable event, we should not expect catastrophic vessel failure at pressures far below the safety factor.

For a major process vessel, we can estimate consequences severity as a simple function of overpressure. Table 5.2 can be applied to Div 1 vessels and B31.3 piping systems.

Overpressure Ratio P / P design	Consequence	Severity Rating
1.0 – 1.2	Process shutdown, gas flaring	1 (asset, production loss)
1.2 – 1.5	Process shutdown, gas flaring	2
1.5 – 2	Potential for gasket leaks	3 (safety)
2 – 3	Vessel damage, major gasket leaks	4 (safety, asset)
3+	Vessel failure	5 (safety, asset)

Table 5.2: Severity Rating vs. Overpressure – ASME Section 8 Div 1 Vessel, B31.3 Piping

The safety factor for Div 2 vessels is 2.4. High pressure vessels are sometimes designed to Div 2 to save cost and weight. Table 5.3 is applicable to Div 2 vessels.

Overpressure Ratio P / P design	Consequence	Severity Rating
1.0 – 1.2	Process shutdown, gas flaring	1 (asset, production loss)
1.2 – 1.5	Process shutdown, gas flaring	2
1.5 – 1.75	Potential for gasket leaks	3 (safety)
1.75 – 2	Vessel damage, major gasket leaks	4 (safety, asset)
2+	Vessel failure	5 (safety, asset)

Table 5.3: Severity Rating vs. Overpressure – ASME Section 8 Div 2 Vessel

Making the Frequency Judgment

Level of Protection Analysis [17] (LOPA) is a semi-quantitative tool for assessing risk. It uses order of magnitude values for initiating event frequency and likelihood of failure of protective layers, to approximate the frequency of occurrence for any given scenario. In rigor, it falls between a typical risk matrix approach (as commonly used in HAZOPs) and a fully quantitative method (QRA). A LOPA is frequently performed after a HAZOP to further investigate significant findings.

The LOPA method is enabled by:

1. Available data on initiating event frequencies (see Tables 5.4 and 5.5).

2. Available data (see Table 5.6).and established calculation methods for probability of failure on demand of IPLs

Initiating Event Frequencies

Initiating Cause	Likelihood, events/yr	Likelihood 10^{-x}
Control Loop Failure	1/10	10^{-1}
Seal failure	1/10	10^{-1}
Gasket failure	1/100	10^{-2}
Unloading hose failure	1/10	10^{-1}
Rotating Equip Failure	1/10	10^{-1}
Rotating Equip Trip	1/1	10^{0}
Fixed Equip Failure	1/100	10^{-2}
Loss of Power	1/10	10^{-1}
Utility Failure	1/10	10^{-1}

Table 5.4 – Initiating Event Frequencies

Human Errors	Likelihood
Well trained operator with stress	1 per10 opportunities
Well trained operator, no stress	1 per 100 opportunities
Well trained operator, no stress, independent verification	1 per 1000 opportunities

Table 5.5 – Human Error Frequencies

Independent Protection Layers

Other non-SIF, independent protection layers (IPLs) also have associated characteristic probabilities of failure on demand (see Table 5.6)

For example, a typical Pressure Safety Valve (PSV) is expected to fail to open in 1/100 to 1/1000 tries. Hence a PSV is assigned a PFD of 0.01 (i.e. it should function correctly for >99% of demands). This corresponds to a risk reduction factor of 2 (orders of magnitude).

IPL	PFD	Risk Reduction
PSV, Rupture disk (clean service, properly sized for the selected scenario)	1/100	2
Independent Control Loop	1/10	1
Open vent, no valve (clean service, properly sized for the selected scenario)	1/100	2
Flame arrestor (prevent flashback) (clean service, properly sized for the selected scenario)	1/10	1
CRO, alarm, well documented action, 20 minutes available	1/10	1

Table 5.6 – Some Typical IPL PFDs

Ignition Probability

Facilities are designed to avoid ignition sources wherever hydrocarbons are contained and might be released. The estimation of ignition probability is beyond the scope of this book. Most gas clouds do not ignite.

Suggestion: Given that a large majority of gas releases do not ignite, it is reasonable in a HAZOP to take a 1 order of magnitude credit on the likelihood that a gas cloud will ignite unless the cloud is very large.

Number of People in the Area

Trevor Kletz famously noted that if no one is around, no one will be killed. Most operating areas are sparsely populated most of the time. Except for scenarios that envelope or risk the entire facility, a limited number of people will be at risk.

An important exception to this is the case of a slowly developing scenario in which outside operators may be directed to the site of an upset.

Suggestion: If there is less than a 10% chance of the area being occupied, then a one order of magnitude credit can be taken.

IPLs vs. Safeguards

HAZOPs identify multiple safeguards. These will not all count as IPLs under LOPA rules. In order to be considered an IPL, a protective function must be:

1. Effective in preventing the consequence when it functions as designed.

    a. Can the IPL detect the condition that requires it to act?

    b. Can it respond in time?

        i. Does the IPL have time to detect the upset, process the information and take the required action (such as valve closing time)? Will this entire process occur rapidly enough to have the desired effect?

    c. Does it have adequate capacity?

2. Independent of the initiating event.

3. Independent of any other IPL for which credit has already been taken.

4. Auditable and testable.

Safeguards Not Usually Considered IPLs

1. Check valves (though 2 dissimilar valves in series may be counted)

2. Procedures, Certification, Training

3. Testing, Inspection, Maintenance

4. Communication Systems, Signs

5. Active Fire Fighting Systems

Determining PFDs of SIFs

Safety Integrity Level (SIL)

The accepted reliability standard for a Safety Instrumented Functions (SIF) is the Safety Integrity level (SIL) rating. This is a measure of the SIF's Probability of Failure on Demand (PFD). Table 6.8 defines SIL levels and associated PFD ranges.

SIL	SIF PFD
1	**1/10** – 1/100
2	**1/100** – 1/1000
3	**1/1000** – 1/10000
4	**1/10000** – 1-1000000

Table 5.7: SIL Definitions

For example, if we require that a SIF work effectively at least 99 times out of 100 tries, then we need a SIL 2 rated SIF.

In order to establish the SIL rating of a SIF, the failure probabilities of the SIF's individual components must be statistically combined and a testing frequency established. The testing frequency is an important consideration. Since these systems are used very infrequently, latent (unrevealed) failure on demand is a concern. Even the best engineered SIF cannot be expected to work forever without maintenance, and its reliability cannot be proved without testing.

The discussion below provides general guidance on what is practically achievable in SIFs.

Achieving SIL 1

A single safety switch actuating a single SDV can generally achieve SIL 1. If a single shutdown switch must actuate several SDVs in parallel (e.g. shutoff multiple feeds) then it will be more difficult to meet SIL 1 (requiring greater component reliability and/or higher testing frequency).

Note: Often the Cause and Effects diagram shows a SIF actuating several different valves or other final elements to achieve a desired process shutdown state. However, closure of a single valve may often achieve the required protective function. In that case, the SIF may be defined as utilizing only the one required valve, making SIL1 easier to achieve.

Achieving SIL 2

Achieving SIL 2 may require multiple sensing devices in a voting arrangement actuating multiple SDVs in series, in conjunction with more onerous testing requirements. Expert guidance is required.

Achieving SIL 3

SIL 3 SIFs require multiple sensing devices in a voting arrangement and multiple SDVs. SIL 3 systems are uncommon in the oil patch. Expert guidance is required.

Achieving SIL 4

It is possible to design a SIL 4 rated SIF.

But if your process requires a SIL 4 SIF, redesign the process to be inherently less hazardous.

CHAPTER 6: SUMMARY OF THE GATE HAZOP PROCESS

So let's recap the GATE HAZOP process:

1. Before the HAZOP begins, the facilitator should:

   a. Organize all P&IDs to be HAZOPed and identify the stream-based nodes.

   b. Generate node sketches for each stream-based. A list of requirements for Node sketches is located in Chapter 3 under Node Sketches.

   c. Print the stream-based node sketches for each team member and attach the corresponding P&IDs to form a packet.

   d. Identify the equipment-based nodes (typical HAZOP nodes).

   e. Determine the appropriate risk matrix for the project.

2. Once the HAZOP is in session, conduct the FLOW discussions for each stream-based node.

   a. For each stream-based node, the responsible design engineer(s) should start by explaining the design and process of the stream.

   b. Then, for each stream-based node, the facilitator should conduct the FLOW deviation discussions. The purpose of the FLOW deviation discussions is largely to develop a detailed understanding of the process.

      i. Where a FLOW deviation causes a TEMPERATURE, PRESSURE or LEVEL deviation, capture the cause in the appropriate equipment-based nodes. Delay discussion of frequency, consequences, and safeguards until the equipment-based node discussion.

   c. Fully discuss other consequences of flow deviation (EG: vibration, erosion, slugging, solids deposition, etc.). A suggested list of deviations is provided in Chapter 3.

3. Once all the stream-based nodes have been discussed and documented, the equipment-based nodes are assessed.

   a. For each equipment-based node, conduct the PRESSURE, TEMPERATURE and LEVEL deviation discussions. Note that at this point almost all causes of pressure, temperature and level deviation will have been identified via the stream-based node discussions.

   b. The discussion now switches to discussion of frequency, consequences and safeguards. Safeguards are discussed only once considering the totality of identified deviations for the equipment-based node.

4. Evaluate risks using the appropriate risk matrix and make recommendations.

5. Following the HAZOP, the Facilitator produces the HAZOP report including a readable section on recommendations/findings and a documentation section that captures the entire HAZOP session discussion.

    a. The design team will evaluate the recommendations and make appropriate changes to the design.

6. It may be necessary to reconvene the HAZOP team to review the changes made to close the HAZOP recommendations.

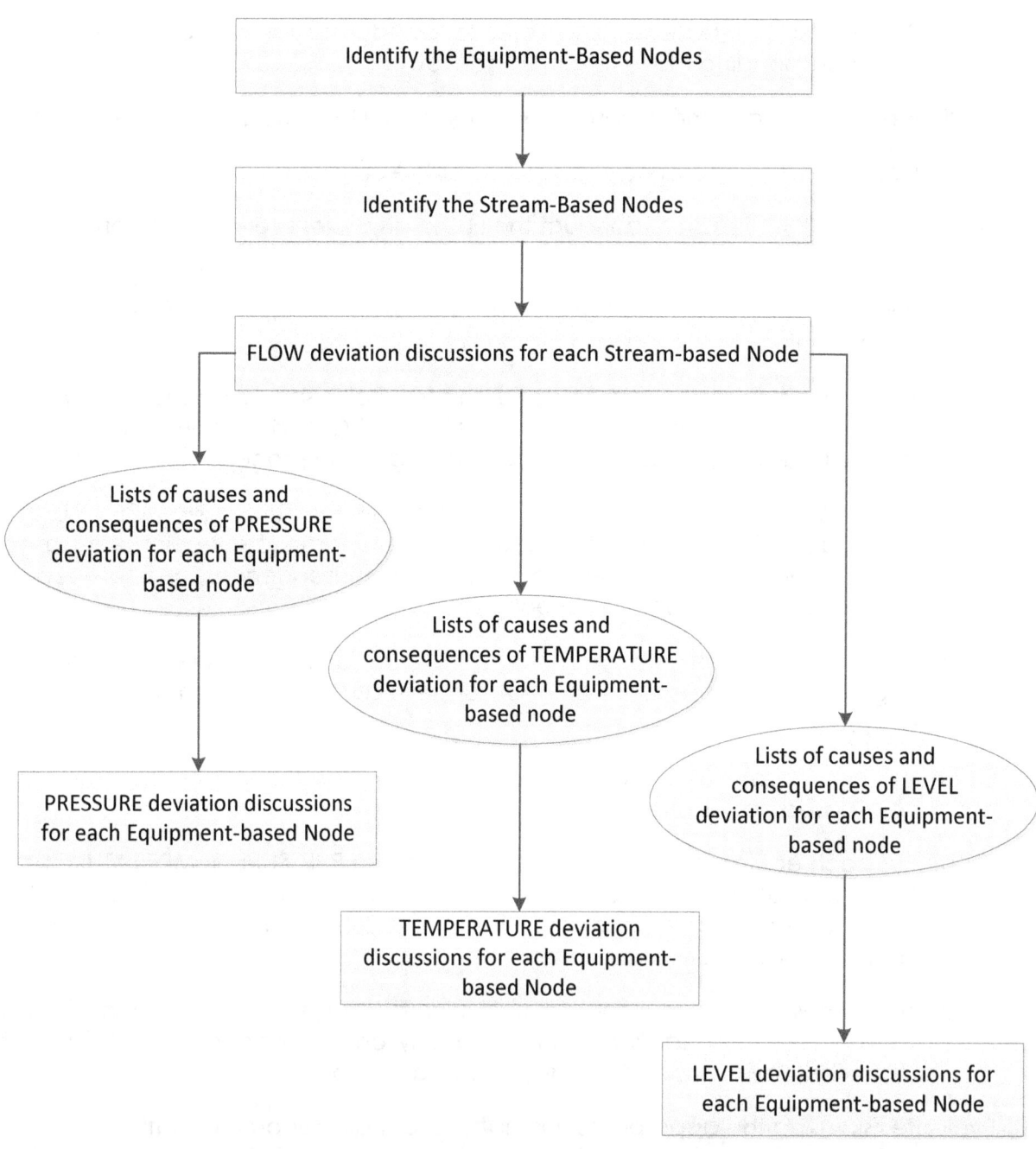

Figure 6.1: GATE HAZOP Method

SECTION II: INTEGRATION WITH OTHER STUDIES

The effectiveness of the HAZOP process can be improved via integration with other studies.

The particular studies selected may vary with the process and the industry. For oil industry projects we recommend:

- API RP 14C Study

- Procedure HAZOP

- Control System Study

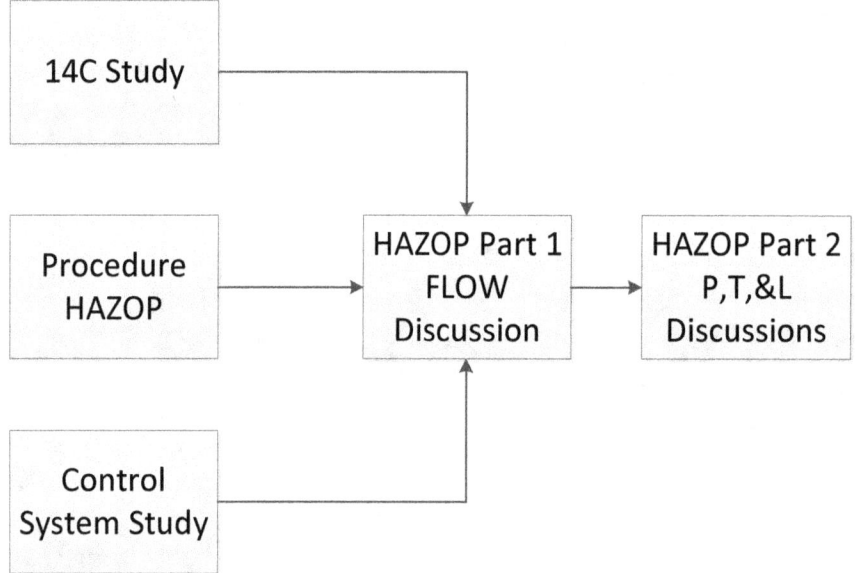

The next two chapters present the recommendations for a Procedure HAZOP and for a Control System Study.

The methodology of a 14C study is well-known and is not discussed.

CHAPTER 7: PROCEDURE HAZOPS

Procedures are not generally considered in a HAZOP. Often they have not been written yet. And even if they have been written, it would be very time consuming to consider detailed procedures with a HAZOP session.

We recommend the following:

1. Write mid-level procedures early in the design process and use them to guide the design process.

2. Conduct a Procedure HAZOP on the mid-level procedures prior to the process design HAZOP.

An operating procedure HAZOP will provide much of the OP frequently missing in a HAZOP.

What is a mid-level procedure?

Procedures may be written at multiple levels of detail. When fully detailed, the procedure for starting a pump may include several steps such as:

- Align manual valves.

- Set control system for startup.

- Inhibit safeguards as required.

- Enable power supply.

- Start pump.

- Reset inhibited safeguards.

- Establish and control forward flow.

The mid-level procedure may simple state: "Start the Pump."

Mid-level steps contain enough information for an experience operator to conduct the step. Mid-level procedures contain enough information to conduct an effective procedure HAZOP.

Procedure HAZOP Method

The review method is fundamentally a two-step process:

1. Describe the planned operation of the system.

2. Brainstorm what could go wrong during that operation.

The brainstorming is structured via a set of questions. For the overall procedure we consider:

- Effectiveness: Will it accomplish the desired objective if implemented as written?

- Completeness: Are any steps missing? (NOTE: It is more difficult to spot omissions than to spot errors. The team should focus dedicated effort on ensuring completeness)

- Order: Are the steps listed in the right order?

- Impact of missing, incomplete, incorrect assumptions or pre-conditions

Following that, we evaluate each mid-level step in the procedure with the aim of answering the question "What could go wrong at this step?" The discussion is structured via the guidewords listed below.

Table 7.1: Questions for Individual steps

Equipment
- What if equipment item fails?
- What if equipment is in unexpected state?

Process Limits
- Pressure: Can the system pressure limit be exceeded?
- Temperature: Can system temperature limits be exceeded?
- Level: What if level limits are exceeded?
- Flow: Is flow induced damage possible? (vibration, erosion, severe slugging)

Human Error
- Are the requirements clear or ambiguous?
- Is the step obvious/common or unusual?
- Is the step simple or complex?
- Does the step extend over a long period of time?
- What if the step is skipped? Is there latent error potential? (skipped step won't be noticed)
- What if the operator does too much or too little?
- What if the operator does this step too early or too late or out of sequence?
- What if the operator does something different?

Other Issues?

Method Basis

The selected study method is based on the

1. Human error research by James Reason[16]

2. Observations from cognitive science research by Fiske and Taylor [9]

3. Typical Procedure HAZOP methodology

Reason[16] has catalogued many causes of human error. The important ones for procedure reviews are summarized in Table 7.2:

	Pre-conditions are likely to be incomplete or not done if or when:
1	Following maintenance/repair the status of equipment may not be as expected. *For instance: slip blinds left in place, equipment not powered, safety systems bypassed, instruments on manual, etc.*
2	Checklist does not follow logical progression. *For instance: valves in a line required to be in a given position should be listed in line order to facilitate checking.*
3	Required conditions prior to proceeding to next step are ambiguous
	Errors in implementing steps in a procedure are most likely:
4	If a step is not obviously cued by the previous step it is likely to be skipped
5	If there are too many steps, steps in the middle are likely to be skipped
6	If a single step is complicated and has sub-steps, slips and lapses are likely
7	A step which extends over long period of time without intervening action is prone to distraction. The operator may not notice cues to continue to the next step. And the operator may forget where he is in the process and may skip steps or repeat previously accomplished steps.
8	Steps which occur after the main goal is achieved are likely to be skipped.
9	Interruptions from SIMOP's (simultaneous operations) can cause you to lose your place.
	Latent Errors – Problems waiting to happen
10	Look for latent errors caused by skipped steps. If a step can be skipped without impacting further steps or causing immediate upset to the process, then it is likely that failure to perform the step will not be caught. Somehow reinforce the need to complete the step.

Table 7.2: Summary of Primary Causes of Human Error

PROCEDURE HAZOP DOCUMENTATION FORM

Procedure: _____

OVERALL PROCEDURE EVALUATION

Effectiveness (Will the procedure as written achieve the desired objectives?)

Completeness (Are any steps or prerequisites missing?)

Order (Are the steps in the right order?)

What if an assumption is incorrect?   What if a condition precedent is incorrect or not achieved?

EVALUATION OF INDIVIDUAL STEPS   (consider each topic for each applicable step in the procedure)

	Discussion/Recommendations
Equipment • Failure • Status	
Process Limits • Pressure • Temperature • Level • Flow • Erosion	
Human Error • Clear/Ambiguous • Obvious/Unusual • Simple/Complex • Long duration • Latent errors • Too much/little • Too Early/late • Different	
Other	

CHAPTER 8: CONTROL SYSTEM REVIEW

'Process Control' is one of the deviations under the FLOW guideword in Table 3.1.

It is arguably the most important discussion point in a HAZOP. Control is the first level of safety. The majority of process risk scenarios are initiated by a loss of control.

Yet most HAZOPs pay only cursory attention to control issues. Within the HAZOP, there is an inherent assumption that we have control unless the control valve either 'fails open' or fails closed'.

Control systems are now very complex and very few people understand the full scope of these systems.

We believe that increased attention to the control system design is required either within the HAZOP session itself, or in a separate study before the HAZOP.

This chapter identifies control system features to be considered in a HAZOP, or preferably in a Control Systems Review conducted prior to the HAZOP.

Control Loop Basics

Figure 8.1 is a simple control loop. It is applicable to both industrial control loops and management control processes.

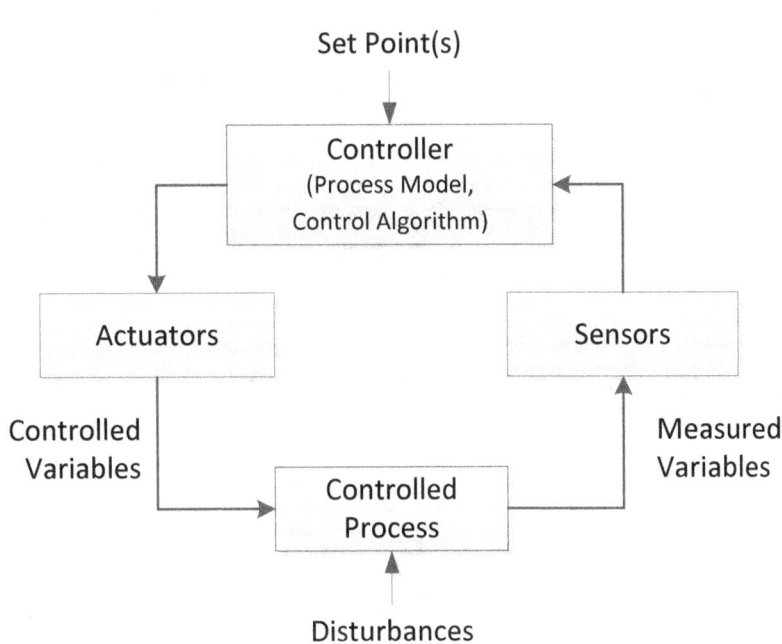

Figure 8.1: Control Loop

There are four required conditions for control:

1. **Goal Condition:** The controller must have a goal. For a simple process control loop the goal is to maintain the set point.

2. **Observability Condition:** Sensor(s) must exist that measure important variable(s). These measurements must provide enough data for the controller to observe the condition of the system.

3. **Model Condition:** The controller must have a model of the system (process model). Data measured by the sensors may be used both to update the model and for direct comparison to the goal or set point. For most loops this is a simple comparison between the measured variable and the set point with controller action defined by a PID algorithm that is adjusted via tuning parameters.

4. **Action Condition:** The actuator must be able to take the action(s) required to achieve the controller goals.

There are multiple opportunities for control loop failure.

Suitable Control Valve

At a very minimum, the HAZOP team should review control valve sizing (review the data sheet). Consider the flowrate and pressure drop design basis.

It is, unfortunately, common practice to size control valves for the highest identifiable flowrate in field life, and to add a 'safety factor', specify a pressure drop much lower than the truth, and then to choose a valve larger than that. The result is often valves that are oversized for any credible operating condition and massively oversized for some operating conditions.

An oversized control valve may not provide control. The HAZOP should catch this.

Process Capacity to Absorb Disturbances

The ability of a loop to keep the process 'between-the-lines' depends to some extent on how far apart the lines are, plus the capacity of the process to absorb system disturbances.

1. Confirm that there is adequate range in the controlled variable to allow for normal variability in the system. For example, when considering pressure in a vessel, the HAZOP team should review:

   - Design Pressure

   - PSV set point (same as design pressure)

   - PSH (high pressure shutdown) (at least 5% below design pressure?)

   - Normal Operating Pressure (at least 5 to 10% below the PSH?)

   - The approximate rate of pressure increase during a credible process disturbance.

The less gap there is between normal operating pressure and shutdown pressure the harder it will be to run the system generally and especially during process upsets.

```
1.00          Design Pressure,  PSV Set Point

0.95          PSHH (High Pressure Shutdown)




O.85          Normal Operating Pressure
```

Figure 8.2: Example – Spread Between Design and Normal Operating Pressures

Failure Modes / Causes

Evaluate fail position of control valves.

If the P&IDs list a failure position (Fail Open, Fail Closed, Fail Last), that generally refers to the failure position on loss of instrument air (or other motive force.) But there are other sources of loop failure. A simple inspection of figure 8.1 suggests these:

1. Sensor failure (high, low, last)

2. Set point change (operator error or remote set point from another controller in a cascade loop)

3. Poor tuning

4. Actuator failure

Sensor Failure

This may be the largest cause of loop failure, especially as sensors become more complex and less well-understood by field personnel.

Some sensing element types are more prone to failure than others. A simple orifice plate flow transmitter is likely to be very robust. If/when it doesn't work someone on the platform will be able to repair it in short order. Some high-tech flowmeters may not work as reliably and, when they fail, there may not be anyone on the platform who knows how to fix it.

When loops depend on sensing elements that are not common, ask how they will fail, what it takes to keep them operating, what skills are required to fix them when they break.

The HAZOP team should enquire about the failure modes of each sensor type as encountered during the study and consider that information in identifying and evaluating risks.

Where safeguarding transmitters exist in addition to control transmitters there is an opportunity for checking. Are the readings compared in the control system and an alarm raised on disparity between the two readings?

Set Point Error

It is common practices to change set points during startup and shutdown and during upset conditions. For example, if a system is to be gradually heated that may be accomplished by gradually increasing the set point on a temperature control loop.

During a process upset, the operator may intentionally change a vessel pressure or level or feed flow, etc. to facilitate recovery.

Set points are sometimes set remotely, as in a cascade loop.

The HAZOP team should identify the risk of inappropriate set points.

Tuning 'Error'

The controller algorithm is an important feature of loop behavior. A poorly tuned loop may oscillate wildly and be itself a hazard to operation.

In a HAZOP we are forced to assume that the loops will be appropriately tuned. Controller behavior is invisible to the HAZOP team.

A preliminary control system study should be performed to identify any loops that will be very difficult to tune.

Control Algorithm

Most industrial control loops are Proportional, Integral, Derivative (PID) loops. A PID controller has three functions:

1. **Proportional action:** Takes action proportional to the error. This is the difference between the measured variable and the set point. Small errors yield minor valve movements, while large errors yield large valve movements.

2. **Integral action:** Takes action proportional to the integral of the error. Here, a small error that has existed for a long time will generate a large valve movement.

3. **Derivative:** Takes action proportional to the derivative of the error. A rapidly changing error generates a large valve movement.

Tuning coefficients are provided for each action type. The appropriate tuning coefficients depend on the dynamics of the process being controlled. The process dynamics can be explained pretty well with three properties, Process Gain, Dead Time and Lag:

- **Process Gain**: is the ratio of measured variable change to control valve position change.

- **Lag:** is a measure of the time it takes the process to get to a new steady state.

- **Dead Time:** is the time between when the valve moves and the process variable <u>begins</u> to change.

Consider the process in place for tuning loops. Many organizations have little or no expertise in this area. Loops that are not tuned cannot be expected to perform effectively.

Actuator Fault

For effective control the actuator must have the capacity to respond at appropriate speed to controller commands.

- What type of actuator was selected?

- Is it correctly specified:

    o Specified for the type of valve installed?

    o Maximum differential pressure?

Cavitation

Cavitation can cause very rapid failure of systems - as little as hours in a serious case.

Cavitation is possible if:

- Single component fluid

- Pressure drop is greater than 50% of the inlet pressure through a single stage.

Complexity

The discussion suggested above can be a part of a larger discussion on complexity. Forty years ago everything one needed to know about a control loop was contained on a loop sheet. Today's loops are often much more complex. And few people can understand the full scope of a loop because much of it is buried in program code.

Control systems are now very complex and several subject matter experts are involved in control system design. The potential for miscommunication resulting dysfunctional control is high. In the rest of this chapter we take a broader systems view of control failure as a source of process risk. Control can fail for reasons other than component failure.

In "Engineering a Safer World" Nancy Leveson[15] argues that:

The process hazard analysis methods we use today were designed for the relatively simple projects of yesterday and are not adequate for the complex projects we build today [16].

She further argues that safety is a control problem including control at multiple levels of hierarchy.

Complex systems exist in a hierarchical arrangement. Even simple rules sets at lower levels of the hierarchy can result in surprising behavior at higher levels. An ant colony is a good example [2]:

A single ant has few skills – a very simple rule set. Alone in the wild it will wander aimlessly and die.

But put a few thousand together and they form a culture. They build and defend nests, find food, divide the work.

CULTURE!? Where did that come from? No scientist could predict ant culture by studying individual ants.

This is the most interesting feature of complex systems. Culture is not contained within individual ants – it is only a property of the collective. This feature is called emergence - the culture emerges.

An emergent property is a property of the network that is not a property of the individual nodes. The sum is more than the parts, and different than the parts.

Safety is Emergent

There is a fundamental problem with equating safety with component reliability. Reliability is a component property. Process safety is emergent. It is a system property.

The system shown in Figure 8.2 is hierarchical. Safety depends on constraints on the behavior of the components in the system including constraints on their potential interactions and constraints imposed by each level of the hierarchy on the lower levels.

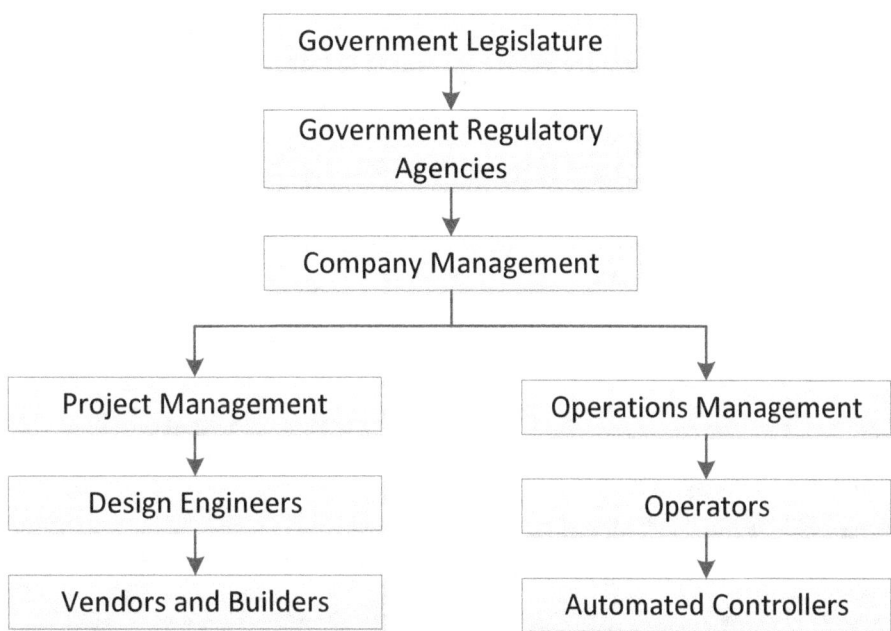

Figure 8.2: Simplified Hierarchy of Project and Operating Asset

Safety as a Control Problem

Safety depends on system constraints. It is a control problem.

And control is required at multiple hierarchical levels. Multiple safeguards at the equipment level are not completely independent if management decisions (such as cutting maintenance funding) tend to degrade all of them.

SECTION III: COGNITIVE AND PSYCHOLOGICAL FACTORS

Chapter 9: Cognitive and Psychological Factors

CHAPTER 9 - COGNIITIVE AND PSYCHOLOGICAL CONSIDERATIONS

Cognitive and Psychological Limits to HAZOP Effectiveness

HAZOPs are successful, in part, because they help us to overcome some cognitive and psychological limits to learning including:

- Anchoring

- Structural secrecy

- Alternative-focused thinking

- Conflicting objectives

- Unactionable plans, groupthink, in-group favoritism

- Lack of motivation to change

- Stress

The stream-based HAZOP approach has notable benefits in the area of structural secrecy at least, and potentially in other areas as well.

This chapter discusses factors that can be harnessed to improve the effectiveness of HAZOPs.

Anchoring [5]

Design teams anchor on past solutions in coming up with a design.

The HAZOP team will be subject to the same tendencies. There will be a tendency to accept familiar designs. For example, designs that meet regulatory requirements for safety systems are usually accepted without serious inquiry even if the objective is to be ALARP.

Recommendations to change the design will be resisted if the design team is firmly anchored on the design.

Anchoring can be especially strong in situations where HAZOP participants have to answer to others not in attendance. If the design is championed by management, a partner, senior members of the design team, etc., then the team member who attended the HAZOP may have a hard time selling the recommendation.

Structural Secrecy [6]

A typical organization is composed of multiple teams of specialists. Each team typically interfaces, formally or informally, with other teams at the same level in the hierarchy while reporting formally up to the next level of management.

Data Collection and Transfer [9]

Individuals collect raw data, evaluate it (add meaning) and then transmit summaries and inferences to other teams. The collection and evaluation of data is not unbiased. Team positions and priorities impact what individual team members believe, what data they collect, how they evaluate the data and what they report. Discussions within teams may be relatively open and honest, but team-to-team and team-to-management communications are likely to be highly filtered and designed more to achieve team objectives than to transmit facts and information. Several factors conspire to dilute the effectiveness of data transfer including:

Ignorance of others needs and knowledge:

If you don't have a good grasp on what members of other groups do and why they do it, you will not be sensitive to their information needs and you will not be sensitive to the impacts that your decisions have on their tasks. You may not pass along the information they need simply because of your ignorance of their needs.

Even when you know that members of other teams need some particular piece of information, you may not provide it if you think they already know it. This is particular true of tacit knowledge. It is very difficult to know what others know and what they don't, especially if the working relationship is distant. Everything you "know" you've inferred from data. Others with the same data may not "know" what you "know" and may well "know" something very different.

Control Issues

Data is power. We all seek control. We cherry pick the data that supports our positions, we evaluate the data to our group's advantage and we pass along this skewed analysis. When data embarrassing to our team is found, it is likely that that data will not be provided to other teams or at least not provided in a timely manner.

Structural Secrecy in HAZOPs

Separate HAZOPs are often done for separate project areas. For instance, the subsea design is usually HAZOPed separately from the topsides design. Where flaws exist because of lack of communication between the two design teams, these separate HAZOPs may not find the problems.

Structural secrecy may also impact the closing of HAZOP recommendations if the party responsible for working the issue does not work effectively with other teams on evaluating the findings/recommendations.

Stream-based Nodes Combat Structural Secrecy

Streams don't recognize team boundaries or contract boundaries.

Some design flaws result from inadequate sharing of data and information between design teams. HAZOPs should identify these flaws, but may not if the HAZOP team doesn't include the right people and/or the information available to the HAZOP team is inadequate.

Separate HAZOPs are often done for separate project areas. For instance, the subsea design is usually HAZOPed separately from the topsides design. Where flaws exist because of lack of communication between the two design teams, these separate HAZOPs may not find the problems.

Very significant structural secrecy problems occur because design engineers are frequently unaware of commissioning, startup and operability issues.

We spend most of our design efforts designing for steady-state operation. Hazards are almost always caused by dynamic events.

A great deal of information needs to be provided to the HAZOP team in a short period of time. Inadequate communication of the design objectives and key features of the design may result in flawed evaluation. This is particularly true in the areas of control systems and instrumented safety systems. The P&IDs do not provide complete information on how these systems will operate.

An interesting sort of structural secrecy is created by the HAZOP feature of considering one node at a time. This tends to shield the participants from the interactions between systems, especially if node boundaries are poorly selected.

Structural secrecy may also impact the closing of HAZOP recommendations if the party responsible for working the issue does not work effectively with other teams on evaluating the recommendation.

Alternative-focused Thinking [7]

Keeney (8) describes our tendency to focus on alternatives rather than objectives. We are action-oriented. Given a problem, we immediately begin trying to identify a solution. Once we identify a promising solution we are predisposed to begin implementing it. We often fail to explicitly identify all the valid objectives during the design process.

If the design team has not explicitly identified the design objectives they will not be able to communicate the design objectives to the HAZOP team. The HAZOP tends to focus attention on a set of design objects related to safety and operability and may be blind to other design objectives not explicitly stated. Also, the HAZOP method tends to focus attention on one objective at a time. The design team has to be able to communicate a comprehensive set of objectives; else some poor recommendations may result.

There will also be a tendency towards alternative-focused thinking during the evaluation of recommendations. The evaluation will likely focus on the features of the problem identified by the HAZOP team rather than evaluating the recommendation against all valid project objectives.

Conflicting Objectives

Design organizations feature many different groups with different objectives. Where these objectives conflict, decisions have to be made. Often the HAZOP raises these issues for the first time. This is especially true if the HAZOP is the first time some stakeholders such as partners and operators get a shot at the design.

Decision making in an environment of conflicting objectives is very demanding. At a minimum the evaluation of recommendations will be time consuming and the eventual decision may be based on political clout rather than what is best for the project.

Conflicting objectives are frequently safety related. It is common for engineering and operations to disagree on 'how safe is safe enough?'

Unactionable Plans, Groupthink, In-Group Favoritism [8,9,10]

In social settings defensiveness impacts the decisions we make and also impacts the reactions of others to those decisions (Argyris9). Defensiveness also impacts the HAZOP session and the evaluation of recommendations. During the session, embarrassing issues may not be raised if the atmosphere is overly adversarial. During evaluation of recommendations the design team will tend to defend their design rather than make changes. This will be especially true if the recommendations come from outside the team; for instance, from partners. Groupthink and in-group favoritism tend to make the design team overly confident of their opinions and hence to resist change (Fisk and Taylor10) (Baron and Kerr11).

If the HAZOP team is undisciplined many of their recommendations may be unactionable. Recommendations are often ambiguous. Recommendations to perform studies or 'consider xxx' can be especially ambiguous and unreasonable and difficult to close.

Systems Effects [13]

Actions frequently result in side effects in addition to their intended effects (Senge14). The now common use of HAZOPs may have the effect of decreasing the design teams focus on safety issues. The author has heard the comment on more than one occasion "if it's a problem the HAZOP will catch it." At least some engineers now believe that they can generate an OK design and then let the HAZOP make it robust.

The Management of Change (MOC) process designed to bring order to projects by controlling and communicating changes also makes it more difficult to correct design flaws.

More complex projects have led to the formation of more complex project organizations featuring more teams and more subject matter experts. There are now more narrowly focused specialists making decisions with global implications, and often making them poorly.

Lack of Motivation to Change [11]

Change occurs only when sufficient motivation exists to overcome the inertia of the status quo (Cialdini12). HAZOPs occur fairly late in the design process when making significant change is painful. Most companies have Management of Change (MOC) processes that intentionally make it difficult to make changes. HAZOP recommendations may be rejected simply because change is too difficult or too late. In any case, significant commitment to change is required to effect the change at this stage of the project. The inertia of the design team will be to reject changes if possible.

Limits to Learning from Failure [8]

Contrary to popular belief we learn mainly from our successes rather than from our failures. Simply identifying a problem is not learning. Learning requires that we also identify a valid solution. The role of the HAZOP is to identify problems; the design team has to solve the problems. In this situation the team is likely to "learn" the least threatening lesson. The least threatening lesson is to disagree with the claim that there is a problem. Failing this the team will tend to identify the least oppressive change that might satisfy the HAZOP action item. If changing operating procedures will suffice, then that change will likely be made rather than changing the design.

Stress and Expertise (Naturalistic Decision Making) [12]

Textbooks would have us believe that we make decisions via this standard pattern:

1. Identify objectives

2. Identify alternatives

3. Evaluate the alternatives to identify the best one

4. Implement the best solution

Most decision making in the real world does not follow that pattern. Engineers use their expertise to short-cut the process. The typical Naturalistic Decision Making (NDM) pattern is (Klein[13]):

1. Evaluate the situation – identify objectives

2. Identify a possible solution

3. Evaluate the solution

4. If the solution is satisfactory implement it.

This process is applied because it is fast and yields satisfactory, if not optimal, solutions. HAZOP teams will tend to apply NDM strategies to identify recommendations. They are limited to this strategy by time limits.

Engineers evaluating the recommendations will also tend to apply an NDM strategy for the same reason; limited time makes exhaustive identification and evaluation of options impractical if not impossible. This can result in tunnel vision. The solution to a given problem may well cause more serious problems elsewhere.

REFERENCES

1. Duhon, Sutton, (2010), "Why We Don't Learn What We Should From HAZOPS", SPE-120735-PA, SPE Projects, Facilities and Construction, June 2010

2. Duhon (2011), "Stream-based HAZOP, A More Effective HAZOP Method", SPE-114982, Presented at SPE Americas E&P Health, Safety, Security and Environmental Conference, Houston, TX, USA, 21-23 March 2011

3. Duhon, Cronin (2015), "Risk Assessment in HAZOPs", SPE-173544-MS, Presented at the SPE Americas E&P Health, Safety, Security and Environmental Conference, Denver, CO, USA, 16-18 March 2015

4. Duhon, (2015), "What's Wrong with HAZOPs and What we can do about it", Prepared for Presentation at American Institute of Chemical Engineers, 2015 Spring Meeting 11th Global Congress on Process Safety, Austin, Texas April 27-29, 2015

5. Khaneman (2011), Thinking Fast and Slow, Farrar, Straus and Giroux

6. Weick (1995) *Sensemaking in Organizations*, SAGE Publications

7. Keeney (1992) *Value-Focused Thinking*, 1992, Harvard University Press

8. Argyris, (2000), *Flawed Advice and the Management Trap*, Oxford University Press

9. Taylor, Fisk, Susan Fisk and Shelley Taylor (1991), Social Cognition, McGraw Hill

10. Baron, Kerr, (2003), *Group Process, Group Decision, Group Action*, Open University Press

11. Cialdini (2001), *Influence, Science and Practice*, Allyn and Bacon

12. Klein (1998), *Sources of Power, How People Make Decisions*, MIT Press

13. Senge (1990) *The Fifth Discipline*, 1990, Currency Doubleday

14. IEC 61822, *Hazard and operability studies (HAZOP Studies) – Application Guide*

15. Leveson, (2011) *Engineering a Safer World, Systems Thinking Applied to Safety*, MIT Press

16. Reason (1990), *Human Error*, Cambridge University Press